Understanding the NEC4 ECC Contract

As usage of the NEC family of contracts continues to grow worldwide, so does the importance of understanding its clauses and nuances to everyone working in the built environment. *Understanding the NEC4 ECC Contract* uses plain English to lead the reader through the NEC4 Engineering and Construction Contract's key features. Chapters cover:

- The Contractor's main responsibilities and the use of early warnings
- Contractor's design
- Tendering
- Quality management
- Payment
- Liabilities and insurance
- Termination
- Avoiding and resolving disputes and much more.

Common problems experienced when using the Engineering and Construction Contract are signalled to the reader throughout, and the correct way of reading each clause is explained. The way the contract affects procurement processes, dispute resolution, project management, and risk management are all addressed in order to direct the user to best practice.

Written for construction professionals, by a practising international construction contract consultant, this handbook is the most straightforward, balanced and practical guide to the NEC4 ECC available. It is an ideal companion for clients, contractors, project managers, supervisors, engineers, architects, quantity surveyors, subcontractors, and anyone else interested in working successfully with the NEC4 ECC.

Kelvin Hughes spent 18 years in commercial management with major contractors, then the past 26 years as a consultant, including a four-year senior lectureship at the University of Glamorgan (now the University of South Wales). He has been a leading authority on the NEC since 1996, was Secretary of the NEC Users' Group for ten years and has run more than 1,600 NEC training courses.

Understanding Construction

Understanding the NEC4 ECC Contract

A Practical Handbook

Kelvin Hughes

Routledge
Taylor & Francis Group

LONDON AND NEW YORK

First published 2019
by Routledge
2 Park Square, Milton Park, Abingdon, Oxon OX14 4RN

and by Routledge
711 Third Avenue, New York, NY 10017

Routledge is an imprint of the Taylor & Francis Group, an informa business

British Library Cataloguing-in-Publication Data
A catalogue record for this book is available from the British Library

Library of Congress Cataloging-in-Publication Data
Names: Hughes, Kelvin (Engineering consultant) author.
Title: Understanding the NEC4 ECC contract : a practical handbook /
    Kelvin Hughes. Other titles: Understanding the new engineering
    contract fourth, engineering and construction contract
Description: New York, NY : Routledge, 2018. | Series: Understanding
    construction | "Routledge is an imprint of the Taylor & Francis
    Group, an Informa business." | Includes bibliographical references and
    index.
Identifiers: LCCN 2018006794| ISBN 9781138499706 (hardback : alk.
    paper) | ISBN 9781138499720 (pbk. : alk. paper) |
    ISBN 9781351014359 (ebook)
Subjects: LCSH: Civil engineering contracts—Great Britain. |
    Construction contracts—Great Britain. | NEC Contracts.
Classification: LCC KD1641 .H8345 2018 | DDC 343.4107/8624—dc23
LC record available at https://lccn.loc.gov/2018006794

ISBN: 978-1-138-49970-6 (hbk)
ISBN: 978-1-138-49972-0 (pbk)
ISBN: 978-1-351-01435-9 (ebk)

Typeset in Goudy Oldstyle
by Swales & Willis Ltd, Exeter, Devon, UK

Contents

Preface

I have been involved with the NEC family of contracts since 1995, and in that time have advised on numerous projects using the family of contracts and carried out over 1,600 NEC-based training courses in the UK and abroad.

I was also Secretary of the NEC Users Group from 1996 to 2006, providing support to users and potential users of the contracts including seminars and workshops, during which time I also ran the Users Group helpline answering queries on the NEC contract from members of the Group.

During this involvement with the NEC, particularly with the Engineering and Construction Contract (ECC), I have always felt there was a need for a practical manual on the contract and its use, including worked examples and illustrations, essentially a training course within a book. For example, what does an activity schedule look like, how do time risk allowances and float work, how do you assess compensation events for omissions of work, etc, which are not covered either by the Guidance Notes, or by other textbooks?

The Guidance Notes that accompany the contract were written by the contract drafters, whom I have known and always admired, and whilst they are comprehensive and well written, I have always felt that they do not adequately act as a practical handbook. Various books have also been developed that have in many cases been written by people who may not be actively involved with NEC contracts and training people to use them.

This book is intended to fill the gap, and to cover issues that may not have been considered or covered previously.

It is intended to be of benefit to experienced professionals who are actually using the contract, but also to students who need some awareness of the contract as part of their studies.

Whilst a number of practitioners still use the NEC3 Engineering and Construction Contract, the book is primarily aimed at giving guidance to NEC4 users, though the structure and content of the contracts are very similar and to that end much of the advice given in this book is of use to all users.

As with the previous books, certain words are capitalised within the text for clarity or emphasis, e.g. Main Options and Secondary Options.

Hopefully readers will not be confused in that, where referring to, say, "a contractor" or "a client" lower case is used, but when referring to "the Contractor" or "the Client" (i.e. the one in the contract), the initial letter is capitalised.

I have also not followed the contract drafting in consistently referring to "the Contractor" as "the *Contractor*" and various other italicised words, as it was felt that this would be confusing to readers, particularly those new to NEC.

I have occasionally mentioned within this book that a particular clause is "a new clause" or "a new provision" within the NEC4 contracts, or in some cases a reversal to how the contract was drafted within or prior to NEC3 (one fee percentage, early warning meetings, etc), but it is not my intention within this book to provide a "clause by clause" comparison of the NEC3 vs NEC4 Engineering and Construction Contract. That has already been done by the contract publishers.

As a general statement, apart from the additional members of the contract family, the NEC4 contracts are not a radical restructure or redraft of the NEC3 contracts, but some fairly minor redrafting to include some additional provisions or to provide additional clarity.

As I have always viewed the NEC contracts as a manual of good practice as well as a contract, particularly in terms of the use of its Main and Secondary Options, risk management through the early warning system, clear requirement for a programme, and a disciplined procedure for change management, as with my previous books on the subject, I have deliberately included within each chapter an overview of the subject area covered by the ECC, and discussed how other contracts deal with the issues, as well as detailing the contract provisions. I hope that will give the readership a more rounded view of good project management principles. Hopefully this will also be useful to student readers.

Readers may note the absence of case law within the text of the book. This is a deliberate policy on my part, for three reasons.

Firstly, I am not a lawyer, my background being in senior commercial positions with major building contractors, so I felt, and readers may concur, that I am not qualified to quote and to attempt a detailed commentary on any case law.

Secondly, there has actually been very little case law on the NEC contracts since they were first launched.

Thirdly, and probably the most important, as a contracts consultant with significant overseas experience of all contracts including the NEC, it was always my intention that the book should attract an international readership. NEC was always conceived as an international contract, so including UK case law would probably limit it to a UK readership.

I have included within the text many examples, as I have found when running NEC training courses that delegates understand principles much better if they can be given illustrated and worked examples, and where numbers are involved using real calculations.

My personal motivation for writing this book is to collate into one volume a significant quantity of material which I have gathered over many years, and to share with others a substantial knowledge and experience of the contract.

Kelvin Hughes
February 2018

Acknowledgements

I would like to extend my love and thanks to the following people who have supported me in life, and in writing this book:

- My wife Lesley, the love of my life, who continues to give me the time, the inspiration and the support to fulfil my life's ambitions;
- My parents, Maureen and Dennis, who made me who I am, and who continue to provide enthusiastic support for what I do. I hope that in return I have made them proud; and
- Michael Hess, friend and colleague, for his keen perception, his wisdom, but most of all, for being one of the nicest people I have ever known and worked with.

Introduction

0.1 Background

The history of the New Engineering Contract stems from 1985, when the Council of the Institution of Civil Engineers (ICE) approved a recommendation from its Legal Affairs Committee "to lead a fundamental review of alternative contract strategies for civil engineering design and construction with the objective of identifying the needs for good practice".

There had been a concern for many years that the policy of construction contracts being sector specific, i.e. building, civil engineering, process, etc and using a single procurement route was outdated. In addition there have been concerns for many years, for example about the time taken to agree final accounts.

In 1986 the specification for the new contract was submitted to the Legal Affairs Committee and in 1991 the first consultation document was issued, with consultation lasting a period of two years. It is perhaps worth mentioning that all of the successive NEC contracts have gone through a consultation process, including the NEC Users Group, so that potential users can offer comments about the proposed new contracts.

The New Engineering Contract (now NEC4), is now being used substantially and by many Clients in the public and private sectors in several countries, most Clients having reported that the contract gives far greater control of time, cost and quality issues combined with greatly improved relationships between the contracting parties.

In July 1994, Sir Michael Latham's report "Constructing the Team" recommended that the New Engineering Contract should be adopted by Clients in both the private and public sectors and suggested that it should become a national standard contract across the whole of engineering and construction work generally. He stated that endlessly refining existing conditions of contract would not solve adversarial problems, that a set of basic principles is required on which modern contracts can be based, and a complete family of interlocking documents is also required.

He stated that the New Engineering Contract (NEC) fulfils many of these principles and requirements, but changes to it would be desirable and the matrix is not yet complete.

Also, public and private sector Clients should begin to use the NEC and phase out their use of "bespoke" documents.

Latham also stated that the most effective form of contract should include:

1 A specific duty for all parties to deal fairly with each other and in an atmosphere of mutual co-operation.
2 Firm duties of teamwork, with shared financial motivation to pursue those objectives.
3 A wholly interrelated package of documents suitable for all projects and any procurement route.
4 Easily comprehensible language, with guidance notes attached.
5 Separation of the roles of contract administrator, project or lead manager and adjudicator.
6 A choice of allocation of risks, to be decided as appropriate to each project.
7 Where variations occur, they should be priced in advance, with provision for independent adjudication if agreement cannot be reached.
8 Provision for assessing interim payments by methods other than monthly valuation, i.e. milestones, activity schedules or payment schedules.
9 Clearly setting out the period within which interim payments should be made, failing which they there should be an automatic right to compensation, involving payment of interest.
10 Providing for secure trust fund routes of payment.
11 While avoiding conflict, providing for speedy dispute resolution by an impartial adjudicator/referee/expert.
12 Providing for incentives for exceptional performance.
13 Making provision where appropriate for advance mobilisation payments.

The NEC family provides for all of Latham's recommendations.

In June 2005, NEC3 was launched as a complete review and update of the whole NEC family, including the introduction of two new contracts, the Term Service Contract and the Framework Contract. Since that date, the Framework Contract and the Supply and Short Supply Contracts have also been introduced into the family.

Alongside the launch of NEC3, the Office of Government Commerce (OGC) announced that after extensive research, they identified NEC3 as the contract which should be adopted by private and public Clients in commissioning their projects.

This was reflected on the inside front cover of the NEC3 contracts when they were launched in June 2005, which stated, "OGC *advises public sector procurers that the form of contract used has to be selected according to the objectives of the project, aiming to satisfy the Achieving Excellence in Construction (AEC) principles. This edition of the NEC (NEC3) complies fully with the AEC principles. OGC recommends the use of NEC3 by public sector construction procurers on their construction projects."*

This note was changed in the 2010 reprint to *"The Construction Clients' Board (formerly Public Sector Construction Clients Forum) recommends that public sector organisations use the NEC3 contracts when procuring construction. Standardising use of this comprehensive suite of contracts should help to deliver efficiencies across the public sector and promote behaviours in line with the principles of Achieving Excellence in Construction."*

In June 2017, NEC4 was launched which is the subject of this book.

Whilst articles written about NEC tend to give prominence to many of the big "flagship" projects carried out using the contracts, of much more significance are the very many more modest size and value projects being procured. It is also in this area that many problems have arisen in the implementation of the contract.

The original objectives of the NEC contracts, and more specifically the Engineering and Construction Contract (ECC), which is the topic of this book, were to make improvements under three headings:

Flexibility

The contract should be able to be used:

- For any engineering and/or construction work containing any or all of the traditional disciplines such as civil engineering, building, electrical and mechanical work, and also process engineering.

  Previously, contracts had been written for use by specific sectors of the industry, e.g. ICE civil engineering contracts, JCT building contracts, IChemE process contracts.
- Whether the Contractor has full, partial or no design responsibility.

  Previously, most contracts had provided for portions of the work to be designed by the Contractor, but with separate design and build versions if the Contractor was to design all or most of the works.
- To provide all the normal current options for types of contract such as lump sum, remeasurement, cost reimbursable, target and management contracts.

  Previously, contracts were written, primarily as either lump sum or remeasurement contracts, so there was no choice of procurement method when using a specific standard form.
- To allocate risks to suit each particular project.

  Previously, contracts were written with risks allocated by the contract drafters, and one had to be expert in contract drafting to amend the conditions to suit each specific project on which it was used.
- Anywhere in the world.

  Previously, contracts included country specific procedures and legislation and therefore either could not be used or had to be amended to be used in other countries. To that end, FIDIC contracts have been seen as the only forms of contract which could be used in a wide range of countries.

To date, NEC contracts have been used for projects as widely diverse as airports, sports stadiums, water treatment works, housing projects, in many parts of the world and even research projects in the Antarctic!

Clarity and simplicity

- The contract is written in ordinary language and in the present tense.
- As far as possible NEC only uses words which are in common use so that it is easily understood, particularly where the user's first language is not English. Previously obscure words such as "whereinbeforesaid", "hereinafter" and "afore-mentioned" were commonplace in contracts. It has few sentences that contain more than forty words and uses bullet points to subdivide longer clauses.
- The number of clauses and the amount of text are also less than in most other standard forms of contract and there is an avoidance of cross-referencing found in more traditional standard forms.
- It is also arranged in a format which allows the user to gain familiarity with its contents, and required actions are defined precisely, thereby reducing the likelihood of disputes.
- Finally, subjective words like "fair" and "reasonable" have been used as little as possible, as they can lead to ambiguity, so more objective words and statements are used.

Some critics of NEC have commented that the "simple language" is actually a disadvantage as certain clauses may lack definition and there are certain recognised words that are commonly used in contracts. There is very little case law in existence with NEC contracts, and whilst adjudications are confidential and unreported, anecdotal evidence suggests that there does not appear to be any more adjudications with NEC contracts than any other, which would tend to suggest that the criticism may be unfounded.

Stimulus to good management

This is perhaps the most important objective of the ECC in that every procedure has been designed so that its implementation should contribute to, rather than detract from, the effective management of the work. In order to be effective in this respect, contracts should motivate the parties to proactively want to manage the outcome of the contract, not just to react to situations. NEC intends the parties to be proactive and not reactive. It does also require the parties and those that represent them to have the necessary experience (sadly often lacking) and to be properly trained so that they understand how the NEC works.

The philosophy is founded on two principles:

- "Foresight applied collaboratively mitigates problems and shrinks risk"
- "Clear division of function and responsibility helps accountability and motivates people to play their part"

Examples of foresight within the ECC are the early warning and compensation event procedures, with the early warning provision requiring the Project Manager and the Contractor each to notify the other upon becoming aware of any matter which could have an impact on price, time or quality.

A view held by many Project Managers is that an early warning is something that the Contractor would give, and is an early notice of a "claim". This is an erroneous view, as firstly, early warnings should be given by either the Project Manager or the Contractor, whoever becomes aware of it first, the process being designed to allow the Project Manager and Contractor to share knowledge of a potential issue before it becomes a problem, and secondly, early warnings should be notified regardless of whose fault the problem is – it is about raising and resolving the problem, not compensating the affected party.

The compensation event procedure requires the Contractor, within three weeks, to submit a quotation showing the time and cost effect of the event. The Project Manager then responds to the quotation within two weeks, enabling the matter to be properly resolved close to the time of the event rather than many months or even years later.

The programme is also an important management document with the contract clearly prescribing what the Contractor must include within its programme and requiring the Project Manager to "buy into it" by formally accepting (or not accepting) the programme. The Project Manager can use the programme to decide the most appropriate way to implement change.

In total the ECC is designed to provide a modern method for Clients, Contractors, Project Managers and others to work collaboratively and to achieve their objectives more consistently than has been possible using other traditional forms of contract. People will be motivated to play their part in collaborative management if it is in their commercial and professional interest to do so.

Uncertainty about what is to be done and the inherent risks can often lead to disputes and confrontation but the ECC clearly allocates risks and the collaborative approach will reduce those risks for all the parties so that uncertainty will not arise.

The Preface to the NEC4 Engineering and Construction Contract states that the key objectives in drafting NEC4 contracts were:

- to provide greater stimulus to good management;
- to support new approaches to procurement which improve contract management; and
- to inspire increased use of NEC in new markets and sectors.

Essential differences between the ECC and other forms of contract:

The ECC is part of a matrix of contracts – Engineering and Construction Contract, Engineering and Construction Short Contract, Term Service Contract, Term Service Short Contract, Term Service Subcontract, Design Build and Operate Contract, Alliance Contract, Supply Contract, Supply Short Contract,

Framework Contract, Engineering and Construction Subcontract, Engineering and Construction Short Subcontract, Professional Service Contract, Professional Service Subcontract, Professional Service Short Contract, Dispute Resolution Service Contract, allowing all parties, whatever the project or service to be provided, to work under similar conditions.

Flexibility of use – the ECC is not sector specific, in terms of it being a building, civil engineering, mechanical engineering, process contract, etc.; it can be used for any form of engineering or construction. This is particularly useful where a major project such as an airport or sports stadium can be a combination of building, civil engineering and major mechanical and electrical elements.

Flexibility of procurement – the ECC's Main and Secondary Options, together with flexibility in terms of Contractor design, allow it to be used for any procurement method whether the Contractor is to design all, none, or part of the works.

Early Warning – the ECC contains express provisions requiring the Contractor and the Project Manager to notify and if required call "an early warning meeting" when either becomes aware of any matter which could affect price, time or quality.

Programme – there is a clear and objective requirement for a detailed programme with method statements and regular updates which provides an essential tool for the parties to manage the project and to notify and manage the effect of any changes, problems, delays, etc.

Compensation events – this procedure requires the Contractor to price the time and "Defined Cost" effect of a change within three weeks and for the Project Manager to respond within two weeks. There is therefore a "rolling" Final Account with early settlement and no later "end of job" claims for delay and/or disruption. It is also more beneficial for the Contractor in terms of its cash flow as the Contractor is paid agreed sums rather than reduced "on account" payments, which are subject to later agreement and payment.

Disputes – the contract encourages better relationships and there is far less tendency for disputes because of its provisions. If a dispute should arise there are clear procedures as to how to deal with it, i.e. adjudication, tribunal.

Mutual trust and co-operation

The first clause of all the NEC contracts (Clause 10.1) has always required the Parties, and their agents, e.g. Project Manager, Supervisor, and now Service Manager, to act as stated in the contract, and to act in a spirit of mutual trust and co-operation.

This mirrors Sir Michael Latham in his report "Constructing the Team" when he recommended that the most effective form of contract should include "a specific duty for all parties to deal fairly with each other and in an atmosphere of mutual co-operation".

This has been slightly changed within the NEC4 contracts in that the former Clause 10.1 has now been split into two separate clauses, Clauses 10.1 and 10.2:

- Clause 10.1: *The Parties, the Project Manager and the Supervisor shall act as stated in the contract.*
- Clause 10.2: *The Parties, the Project Manager and the Supervisor act in a spirit of mutual trust and co-operation.*

If we examine the clauses, firstly and rather curiously, Clause 10.1 is written in the future tense, which is unusual for NEC contracts in that they are written in the present tense. This notwithstanding, is it necessary to state that the *Parties, the Project Manager and the Supervisor* are required to act as stated in the contract? What difference does it make if that clause was absent or deleted? Would they not have to act as stated in the contract?

Secondly, what does it mean that the Parties, the Project Manager and the Supervisor are required to act in a spirit of mutual trust and co-operation?

This second requirement mirrors Sir Michael Latham in his report "Constructing the Team" when he recommended that the most effective forms of contract should include "*a specific duty for all parties to deal fairly with each other and in an atmosphere of mutual co-operation*".

This clause has often been viewed with some confusion, and for those who have spent many years in the construction industry, with a degree of scepticism. Most practitioners state that their understanding of the clause is that the parties should be non-adversarial toward each other, acting in a collaborative way and working for each other rather than against each other, and in reality that is what the clause requires.

However, the difficulty is, that if a party does not act in a spirit of mutual trust and co-operation what can another party do? The answer is that the clause is almost unenforceable as it is virtually impossible to define and quantify the breach or the ensuing damages that flow from the breach.

In addition, the Contractor is not contractually related to the Project Manager or Supervisor, so either would be unable to take action directly against the other for breach of contract other than through the Client.

To that end, as Clause 10.1 is a fairly redundant clause, and Clause 10.2 is unenforceable, Clients have been seen to insert a Z clause to delete the requirements; however, it is recommended that it should remain in the contract, if merely viewed as a statement of good intent.

In effect, a clause requiring parties to act in a certain spirit will probably not, on its own, have any real effect. Within the ECC it is the clauses that follow within the contract which require early warnings, clearly detailed programmes which are submitted for acceptance, and a structured change management process, that actually create and develop that level of mutual trust and co-operation rather than simply inserting a statement within the contract requiring the parties to do so.

0.2 Structure of the ECC

The ECC includes the following sections:

Core clauses

1   General
2   The Contractor's main responsibilities
3   Time
4   Quality management
5   Payment
6   Compensation events
7   Title
8   Liabilities and insurance
9   Termination

Main option clauses

Option A   Priced contract with activity schedule

Option B   Priced contract with bill of quantities

Option C   Target contract with activity schedule

Option D   Target contract with bill of quantities

Option E   Cost reimbursable contract

Option F   Management contract

Dispute resolution

Option W1   Used when Adjudication is the method of dispute resolution and the United Kingdom Housing Grants, Construction and Regeneration Act 1996* does not apply

Option W2   Used when Adjudication is the method of dispute resolution and the United Kingdom Housing Grants, Construction and Regeneration Act 1996* applies

*See amendment regarding the Local Democracy, Economic Development and Construction Act 2009.

Option W3   Used when a Dispute Avoidance Board is the method of dispute resolution and the United Kingdom Housing Grants, Construction and Regeneration Act 1996* does not apply

Secondary option clauses

Option X1   Price adjustment for inflation

Option X2   Changes in the law

Option X3   Multiple currencies

Option X4   Ultimate holding company guarantee

Option X5   Sectional Completion

Option X6   Bonus for early Completion

Option X7   Delay damages

Option X8   Undertakings to the Client or Others

Option X9   Transfer of rights

Option X10   Information Modelling

Option X11   Termination by the Client

Option X12   Multiparty collaboration

Option X13   Performance bond

Option X14   Advanced payment to the Contractor

Option X15   The Contractor's design

Option X16   Retention

Option X17   Low performance damages

Option X18   Limitation of liability

Option X20   Key Performance Indicators

Option X21   Whole life cost

Option X22   Early Contractor involvement

Option Y(UK)1   Project Bank Account

Option Y(UK)2   Housing Grants, Construction & Regeneration Act 1996

Option Y(UK)3   Contracts (Rights of Third Parties) Act 1999

Option Z   Additional conditions of contract

N.B. Option X19 is not used

Schedules of Cost Components

Contract Data

Other documents to be used with the ECC include

- the Scope
- the Site Information
- the Accepted Programme
- other documents resulting from choosing various Secondary Options, e.g. Performance Bond

Depending on the choice of Main Options the documents may also include:

- an activity schedule (Option A or C), or
- a bill of quantities (Option B or D)

The Main Options

The six Main Options A to F enable Clients to select a procurement strategy and payment mechanism most appropriate to the project and the various risks involved. Chapter 5 includes more details on payment mechanisms and requirements under each Main Option.

Option A   Priced contract with activity schedule

Option B   Priced contract with bill of quantities

Option C   Target contract with activity schedule

Option D   Target contract with bill of quantities

Option E   Cost reimbursable contract

Option F   Management contract

- Options A and B are priced contracts in which the risks of being able to carry out the work at the agreed prices are largely borne by the Contractor.
- Options C and D are target contracts in which the Client and Contractor share the financial risks in an agreed proportion.
- Options E and F are two types of cost reimbursable contract in which the financial risks of being able to carry out the work are largely borne by the Client.

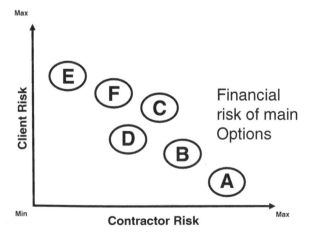

Figure 0.1 Financial risk of Main Options

Essentially, the Main Options differ in the way the Contractor is paid. Whilst many traditional contracts are based on bills of quantities, there has been a movement away from the use of traditional bills and towards payment arrangements such as milestone payments and activity schedules, with payment based on progress achieved, rather than quantity of work done. There is also an increasing use of target cost contracts which has been encouraged by the increasing use of partnering arrangements, and the better sharing of risk.

To that end, it is perhaps not surprising that in surveys carried out by the RICS (Contract in Use Surveys) Options A and C were found to be the most regularly used NEC3 Main Options, with Option B in third place, but significantly behind Options A and C, and Options D, E and F in the "very rarely used" category. It is likely that this trend will continue as use of the NEC4 ECC continues.

Once the procurement strategy has been decided, the Main and Secondary Options can be selected to suit that strategy.

Option A – Priced contract with activity schedule

Procurement using specification and drawings or activity schedules is still the way the largest number of projects, particularly those lower in value, are procured. In addition, many projects are procured on a design and build basis. In that respect, Option A has always been the most commonly selected Main Option.

Option A is normally used where the Client knows exactly what it wants, and is able to clearly define it through the Scope, which would comprise drawings and specifications, but can alternatively be a performance specification, where the Contractor is to design the works to meet specific performance objectives.

The Scope must be sufficiently defined for the Contractor to commit itself to a lump sum price and prepare the Activity Schedule which will identify its payments.

Whilst the ECC is appropriate for all forms of engineering and construction, Option A is particularly appropriate for building projects where the work can more often be clearly defined.

The Activity Schedule is a list of activities, normally prepared by the Contractor and can be used where the Client has provided all the Scope, or where the Contractor is to design all or part of the works based on some form of performance criteria, and therefore there is Scope provided by the Contractor.

An Activity Schedule is a list of activities prepared by the Contractor which it expects to carry out in Providing the Works. When the Contractor has priced it, the lump sum for each activity is the Price to be paid by the Client in the assessment following completion of that activity.

Option A is therefore a stage payment contract and as payment is linked to completion of activities, the Contractor must plan and carry out its work effectively with the cash flow requirements for both parties being clearly visible. Administration of payment aspects is therefore fairly simple.

Establishing and taking off the quantities of work involved to achieve the completion of each activity is the responsibility of the tendering Contractors. The price for each activity is in effect a lump sum for that activity and must include everything necessary to complete the activity. The sum of the tendered lump sums for each of the activities is the tendered prices (the total of the Prices) for the whole of the Works.

Option A is ideally suited to Contractor's design but can be used for Client's design or split design responsibility.

Under Option A, the Contractor prepares and submits an Activity Schedule with its tender. The total of the Prices in the Activity Schedule is the Contractor's price for Providing the Works. The contract requires that the activities on the Activity Schedule relate to operations on each programme.

It is important to recognise that the Activity Schedule has two primary functions:

- it shows how the tendering Contractors have built up their price and can therefore be used as part of the tender assessment process.
- it is used to calculate the Price for Work Done to Date.

The price for each activity is in effect a lump sum for carrying out and completing that activity and must therefore include everything necessary to complete that activity.

The Client does not provide a bill of quantities with Option A, the taking-off of the quantities of work involved to achieve the completion of each activity and the pricing activity being the responsibility of the tendering Contractors.

There are certain advantages to the Client with using Option A:

- there is no requirement for a bill of quantities to be prepared for tendering Contractors to price. This will save time and money at the pre-tender stage.
- the Contractor holds the risk of inaccuracies in the quantities, or of missing an item of work stated within the Scope.
- the assessment of the Price for Work Done to Date is easier and quicker than with the other Main Options. There is no requirement to remeasure the work and apply quantities to rates and prices to calculate the amount due to the Contractor.
- programme and activity schedule preparation are linked as integrated activities which would tend to lead to a more comprehensive tender.
- payment is linked to completion of an activity or group of activities related back to the programme, so cash flow requirements for both parties are more visible.
- in order to receive payment, the Contractor has to complete an activity by the assessment date, so it has to price and programme realistically and is motivated to keep to that programme during construction.
- the assessment of the effects of compensation events is related back to the activity schedules. Any change in resources or methods associated with an activity can be compared with those stated in the Accepted Programme before the compensation event occurred.

The possible disadvantages to the Client are:

- as there is no specific document that all the tendering Contractors price to, it is difficult to assess tenders on a "like for like" basis. Some clients cite that as an advantage, saying that they do not require the facility to check tenders on a line by line basis anyway, so that may be considered as an advantage!

  It is not uncommon, to assist the assessment of tenders, for Clients to issue templates for the Contractor to price covering the main construction elements as part of the tender, with the tendering Contractors then inserting the activities within each element. In that way, each element will have been priced on a "like for like" basis, but the activities within each element may have been described and priced differently by each tenderer.

  If a template is used, Clients should be wary of them being too prescriptive and almost being seen by tenderers as a bill of quantities.
- The tendering Contractors have to spend time, and money, in taking off the quantities of work in order to establish and populate the activity schedule. In order to reduce the tenderers' efforts in producing their own quantities, some Clients issue quantities to tenders, but it must be stressed that these are for indicative purposes only, and tenderers should not rely on those quantities.
- The Client holds the risk of clearly defining the Scope in order that the Contractor can prepare the Activity Schedule.

The author has seen many "hybrids" of Option A contracts (and all the other Main Options), for example, as stated above, Clients preparing an outline activity schedule in the form of a template for tenderers to develop and submit with their tenders.

When using Option A, tenderers should not be restricted in terms of the structure of their activity schedule. Option A gives the Contractor the greatest risk of all the Main Options in terms of pricing tenders, but also in terms of setting up their cash flow requirements, and therefore as far as possible, "hybrid" versions of Option A which restrict pricing and cash flow should be discouraged.

Option A, and the other Main Options are discussed further within this book, primarily within Chapter 5 (Payment).

Option B – Priced contract with bill of quantities

Option B is essentially a remeasurement contract based on a Bill of Quantities prepared by the Client with the Contractor pricing the items in the bill of quantities including matters which are at its own risk. The Contractor is then paid for the quantity completed to date at the rates and prices in the bill of quantities.

In the early days of the New Engineering Contract, Clients would tend to select Option B initially as it mirrored the traditional bills of quantities-based contracts they were using at the time. Once Clients became more experienced in the use of the contract, they tended to move toward other Main Options.

It is important that, if selecting a remeasurement contract, one should consider the merits of preparing and using bills of quantities against other procurement methods.

A Client should not select remeasurement just because in the past, with other forms of contract, it has always begun the procurement process by producing the drawings and specification then measuring the work! Again this comes back to a procurement strategy consideration.

It must be noted that changes which are compensation events are inserted into the bill of quantities as lump sums rather than on a remeasurement basis (see Chapter 6).

This option is normally used where the Client knows what it wants and is able to clearly define it through the Scope, and measure it within the Bills of Quantities, but there are likely to be changes in the quantities, which may or may not be considered as compensation events.

Although there has been a general decline in the use of bills of quantities in the past 25 years, the option is still quite widely used on civil engineering projects where the final quantity of excavations, filling materials, etc. cannot be reasonably forecast.

Option B does not work well when there is a significant amount of Contractor's design; in this case, a lump sum contract such as Option A or perhaps Option C should be used rather than remeasurement, as the Contractor can then carry out initial design calculations, price the work and include the various design stages within its activity schedule rather than have the Client measuring the work, which would not be practical.

A bill of quantities comprises a list of work items and quantities prepared by the Client and priced by the Contractor. Standard methods of measurement are published, which state the items to be included and how the quantities are calculated.

Confusion can sometimes arise as to whether the tendering Contractors should price the Bills of Quantities or the Scope. The Contractor Provides the Works in accordance with the Scope (Clause 20.1), therefore it is assumed that it programmes the Works in accordance with the Scope. Clause 31.2 (Programme) refers to "Providing the Works" and "Scope".

Information in the Bill of Quantities is not Scope or Site Information (Clause 56.1). It does not tell the Contractor what it has to do, but it must clearly be a fair representation of the scope of the Works to allow the Contractor to accurately price the Works. Although a Contractor will often use the quantities within the Bill of Quantities as a guide to calculate time scales, the bill is used as a basis for inviting and assessing tenders, so it is a pricing and payment document, i.e. it deals with money. It is reasonable for the Contractor to presume that the quantities are a fair representation of what is included in the Scope.

If there is an inconsistency between the documents which form part of the contract either the Project Manager or the Contractor notifies the other as soon as either becomes aware that it exists (Clause 17.1) and the Project Manager corrects it. This may then be addressed as a compensation event under Clause 60.6, with Clause 60.7 stating that in assessing a compensation event resulting from the correction of an inconsistency between the Bill of Quantities and another document, the Contractor is assumed to have taken the Bill of Quantities as correct. This is in respect of the Contractor's price, Completion Date and programme, so the Price and, if appropriate, the Completion Date may then be changed to accommodate the correction.

Clearly, if the Contractor notices the error during the tender period then it should raise it as a query, then the matter can be dealt with and all tenderers informed, however, while the Contractor should notify if and when it finds an error, it is not under any obligation to look for the error!

Option C – Target contract with activity schedule

Option C is a target contract based on an activity schedule, again, as with Option A, normally prepared by the Contractor. It must be emphasised that the activity schedule is not used for payments in the same way as with an Option A contract. In Option C, the activity schedule acts as a guide when assessing tenders, but payments are based on Defined Cost incurred, not completion of activities.

Target contracts are a development of cost reimbursable contracts and are a combination of a lump sum price and open-book cost reimbursable payments.

The advantages of target contracts are that the parties initially have certainty of price, with the Contractor being incentivised to make cost savings for the benefit of the Client and itself.

This option is normally used where the Client knows what it wants and is able to clearly define it through the Scope, so that the Contractor can price and prepare the activity schedule but it sees a benefit in sharing risk and opportunity with the Contractor, thereby encouraging collaboration. Financial risks are shared, proportionally between the Client and the Contractor through the Contractor's share percentages. Many wrongly call Option C the "Partnering Option", although it does lend itself to the principles of partnering, collaborative working and associated risk/opportunity sharing. Coincidentally, there is a Partnering (multiparty collaboration) Secondary Option X12 which may have given rise to the confusion. This option is described in detail later within this chapter.

The Contractor tenders a price and includes an Activity Schedule in the same way as it would under an Option A contract. This price, when accepted is then referred to as the "target". The Contractor also tenders its percentage for Fee. The original target is referred to as the Total of the Prices at the Contract Date.

- the target price includes the Contractor's estimate of Defined Cost plus other costs, overheads and profit to be covered by its Fee.
- the Contractor tenders its Fee in terms of a percentage to be applied to Defined Cost.
- during the course of the contract, the Contractor is paid Defined Cost plus the Fee.
- the target is adjusted for compensation events and also for inflation (if Option X1 is used).

- on Completion, the Project Manager assesses the Contractor's share in accordance with Clause 54.1 which, it has to be said, is at best a confusing clause, though the Guidance Notes clarify the clause! The Contractor then pays or is paid its share of the difference between the final total of the Prices and the final Payment for Work Done to Date according to a formula stated in the Contract Data. This motivates the Contractor to decrease costs. Many refer to this sharing of risk and opportunity as "pain and gain". It often comes as a surprise to NEC users that the terms "pain and gain" do not appear anywhere within the NEC contracts, nor does the term "target price"!

The Contractor is required to submit forecasts at the intervals stated in the Contract Data of the Total of the Prices from the starting date to the Completion of the whole of the works, which advises the Client of its potential outturn cost.

The Contractor is also required to advise the Project Manager on the practical implications of the design of the works and on subcontracting arrangements.

Option D – target contract with bill of quantities

Option D is a target contract based on a Bill of Quantities. Again, as with Option B, the Bills of Quantities are prepared by the Client.

This option is normally used where the Client knows what it wants and is able to clearly define it through the Scope, and measure it within the Bills of Quantities, but there are likely to be changes in the quantities, which may or may not be considered as compensation events.

The Contractor tenders a price based on the Bills of Quantities. This price, when accepted is then referred to as the "target". The original target is referred to as the Total of the Prices at the Contract Date.

The assessment of payments is again the same as for an Option C contract:

- the target price includes the Contractor's estimate of Defined Cost plus other costs, overheads and profit to be covered by its Fee.
- the Contractor tenders its Fee in terms of a percentage to be applied to Defined Cost.
- during the course of the contract, the Contractor is paid Defined Cost plus the Fee.
- the target is adjusted for compensation events and also for inflation (if Option X1 is used).
- on Completion, the Contractor is paid (or pays) its share of the difference between the final total of the Prices and the final Payment for Work Done to Date according to a formula stated in the Contract Data. If the final Payment for Work Done to Date is greater than the final total of the Prices, the Contractor pays its share of the difference.
- as with Option B, the target is generated as a remeasurement, based on the Bill of Quantities, though changes which are compensation events are

inserted into the Bill of Quantities as lump sums rather than on a remeasurement basis (see Chapter 6).
- the target is adjusted for compensation events and also for inflation (if Option X1 is used).

Again, as with Option C, the Contractor is required to submit forecasts at the intervals stated in the Contract Data of the Total of the Prices from the starting date to the Completion of the whole of the works.

The Contractor is also required to advise the Project Manager on the practical implications of the design of the works and on subcontracting arrangements.

Option E – cost reimbursable contract

Option E is a cost reimbursable contract with the Contractor being reimbursed Defined Cost plus the Fee.

It should be used:

- where the scope of work is uncertain, e.g. some refurbishment projects;
- where extreme flexibility is required, e.g. for enabling work;
- where a high level of Client involvement is envisaged;
- for emergency work;
- where trials or work of an experimental nature are carried out.

The option allows development of the design as the works proceed and permits maximum flexibility in allocation of design responsibility.

A cost reimbursable contract should be used where the definition of the work to be done is inadequate even as a basis for a target price and yet an early start is required. In such circumstances the Contractor cannot be expected to take risks. It carries minimum risk and is reimbursed its Defined Cost plus Fee, subject only to a number of constraints designed to motivate efficient working.

A criticism of cost reimbursable contracts such as Option E is that it gives the Contractor very little incentive to reduce costs, however a cost reimbursable contract should be used where the definition of the work to be done is inadequate for the Contractor to price and yet an early start is required. In such circumstances the Contractor cannot be expected to take financial risks. It carries minimum risk and is reimbursed its Defined Cost plus Fee, subject to any deduction for Disallowed Costs.

Under Option E of the ECC, the Contractor is reimbursed its Defined Cost plus a Fee, basically covering its offsite overheads and profit. This Fee is calculated by applying the fee percentage, given at tender by the Contractor in Contract Data Part 2, to appropriate Defined Cost. Although Option E is a cost reimbursable contract, there is an obligation on the Contractor to provide a regular forecast of the Total of the Prices which advises the Client of its potential outturn cost.

Another criticism of Option E is that the tendering Contractors do not actually price the works; they just price their fee percentage together with any appropriate Equipment rates, manufacture and fabrication rates and overheads percentages. In addition tenderers may be required to notionally agree to a cost plan. This makes it difficult to assess cost reimbursable tenders, but again the basis of cost reimbursable contracts is that the Client bears most of the risk.

The Contractor is required to submit forecasts at the intervals stated in the Contract Data of the Total of the Prices from the starting date to the Completion of the whole of the works.

The Contractor is also required to advise the Project Manager on the practical implications of the design of the works and on subcontracting arrangements.

Option F – Management contract

Management contracting emerged as a procurement method for large projects in the early 1980s, but its use declined in the 1990s and it is now rarely used, but the ECC provides for it through Option F.

The principle with a management contract is that the Contractor focuses on managing the contract rather than physically building the project and does not normally carry out any construction work itself.

The Contractor's responsibilities under a management contract normally cover:

(i)   provision of site accommodation, and other common use facilities.
(ii) managing time-based deliverables such as programme, milestones and project completion.
(iii) management of cost, normally in conjunction with the Client's cost consultant.
(iv) management of quality and defects related issues.
(v)   managing the design, though in most cases, it is not responsible for carrying out the design – "design and manage" is a procurement method where the Contractor normally has full design responsibility.
(vi) assembling, tendering and managing subcontractor packages.

The amounts paid to the Subcontractors are reimbursed to the Contractor, plus a management fee.

The Contractor holds very little risk under this procurement method, as apart from any proven negligence on its part, it is reimbursed in full the subcontractor costs plus its fee. Any further costs incurred, which cannot be recovered from a Subcontractor, in respect of say, subcontractor insolvency, are reimbursed by the Client.

The Contractor's work mainly applies to the construction phase though it can be appointed for pre-construction services. All subcontracts are directly with the Contractor.

The ECC specifically states under Option F that the Contractor's responsibility is to manage the design that has been assigned to him, the provision of site services and the construction and installation of the works, all of which are subcontracted. The Contractor is also required to advise the Project Manager on the practical implications of the design of the works and on subcontracting arrangements. It is also required to prepare forecasts of the total Defined Cost for the whole of the works in consultation with the Project Manager at intervals stated in the Contract Data.

The Contractor tenders a fee percentage and its estimated total of the prices of the subcontracts. The subcontract prices plus any work carried out by the Contractor itself are then paid to the Contractor as Defined Cost plus Fee. The Contractor is responsible for supplying management services and completing or advising on design if required. If the Contractor is responsible for design it will be appointed on a design and manage basis.

Management contracts are generally suitable:

- where there is a need to co-ordinate a number of works contractors and suppliers;
- when the Client does not have sufficient capability to manage the project;
- when the time scale of the project is tight, requiring an early start of construction.

At this point the scope of the project is not fully developed. As the scope is developed and construction progresses, successive works contracts can be awarded, but the interfaces between these successive packages must be managed.

Again, as with Option E, the tendering Contractors do not actually price the works, they just price their fee percentage, with further rates or percentages as with the other options as the Schedule of Cost Components does not apply. Again, this makes it difficult to assess management contract tenders, but again the basis of this type of contract is that the Client bears most of the risk.

As with Options C, D and E, the Contractor is required to submit forecasts at the intervals stated in the Contract Data of the Total of the Prices from the starting date to the Completion of the whole of the works.

The Contractor is also required to manage the Contractor's design, provision of site services and the construction of the works. The Contractor subcontracts all work except work which the Contract Data states that it will do.

Secondary Options

Note that Option X19 is not used, but as NEC4 has a common numbering system for Main and Secondary Options, any missing Options are found in other members of the NEC4 family, e.g. Option X19 is in the NEC4 Term Service Contract.

This common numbering for the Main and Secondary Options provides a consistency sadly lacking in many other families of standard contracts.

So, in NEC4, X1 will always be "Price adjustment for inflation"; that option will be X1 and the wording will be almost identical.

This feature makes the NEC4 contracts very user friendly, particularly for those who use a number of members of the family and more than one procurement strategy.

Option X1: Price adjustment for inflation (used only with Options A, B, C and D)

The Client should make the decision at the time of preparing the tender documents as to whether inflation for the duration of the contract is to be:

- The Contractor's risk – in which case it should not select Option X1.
- The Client's risk – in which case, it should select Option X1.

The default within the ECC is that the contract is "fixed price" in terms of inflation, i.e. the Contractor has priced the work to include any inflation it may encounter during the period of carrying it. If Option X1 is chosen, the Prices are adjusted for inflation as the work progresses, by means of a formula.

Note that under Options C and D, whether Option X1 is selected or not, the Price for Work Done to Date is the current cost at the time it is incurred. Option X1 is then applied to the Total of the Prices (the target).

Option X1 is not applicable to Options E and F as the Client again pays Defined Cost at the time that it is incurred.

The key components of the formula are:

- The "Base Date Index" (B) is the latest available index before the Base Date.
- The "Latest Index" (L) is the latest available index before the assessment date of an amount due.
- The "Price Adjustment Factor" is the total of the products of each of the proportions stated in the Contract Data multiplied by $(L - B)/B$ for the index linked to it.

Under Options A and B, the amount due includes an amount for price adjustment which is the sum of:

- the change in the Price for Work Done to Date since the last assessment of the amount due multiplied by the PAF, and
- the amount for price adjustment included in the previous amount due.

The change in the Price for Work Done to Date = £50,000
The Base Date Index (B) = 280.0
The Latest Index (L) = 295.5
The Price Adjustment Factor is therefore $(L - B)/B$
$$= (295.4 - 280.0)/280.0$$
$$= 0.055$$
Inflation since the base date is therefore 5.5%
The amount due in this assessment is therefore
£50,000.00 × 0.055 = £2,750.00

Under Options C and D, the amount due includes an amount for price adjustment which is the sum of:

- the change in the Price for Work Done to Date since the last assessment of the amount due multiplied by (PAF/(1+PAF)) where PAF is the Price Adjustment Factor for the date of the current assessment, and
- correcting amounts, not included elsewhere, which arise from changes to indices used for assessing previous amounts for price adjustment.

This amount is then added to the Total of the Prices, i.e. the target.

Note that if Option X1 is chosen, then Defined Cost for compensation events is assessed by adjusting current Defined Cost back to the base date.

Note that for compensation events the Defined Cost is assessed using

- the Defined Cost at base date levels for amounts in the Contract Data for people and Equipment, and
- the Defined Cost current at the dividing date used in assessing the compensation events adjusted to the base date by dividing by "one plus the PAF" for the last assessment due before that dividing date for other amounts.

Aside from Clients favouring fixed price contracts, the rules of Option X1 are probably one of the reasons why this Option is rarely chosen!

Option X2: Changes in the law

As with Option X1, the default is that the contract is "fixed price" in terms of changes in the law, i.e. the Contractor has priced the work to include any changes in the law it may encounter during the period of the contract.

If Option X2 is chosen, and a change in the law occurs after the Contract Date the Project Manager notifies the Contractor of a compensation event. The Prices may be increased or reduced in addition to providing for any delay to Completion.

Note that Option X2 refers to a change in the law of the country in which the Site is located, so for example, a change in the law in another country where goods are being fabricated for delivery to the Site will not be a compensation event.

Option X3: Multiple currencies (used only with Options A and B)

The currency of the contract is stated in Contract Data Part 1.

Option X3 provides for items or activities to be paid in an alternative currency, the items or activities, the currency and the total maximum payment in this currency to be listed in the Contract Data, beyond which payments are made in the currency of the contract.

The exchange rates, their source and date of publication are also referred to in the Contract Data.

Option X4: Ultimate holding company guarantee

This Option was referred to in the NEC3 contracts as Parent Company Guarantee.

This form of guarantee is given by an ultimate holding company (parent company), to guarantee the proper performance of a contract by one of its subsidiaries (the Contractor), who whilst having limited financial resources itself, may be owned by a larger financially sound parent company. In most cases it is the ultimate parent company that provides the guarantee, but sometimes, particularly when the ultimate parent company is in another country, the parent company may just be a company further up the chain within the group, perhaps the national parent who has sufficient assets to provide the required guarantee.

If a parent company guarantee is required, it may either be provided by the Contract Date or within four weeks of the Contract Date, and must be in the form stated in the Scope.

The parent company guarantee should have an expiry date, which could be completion of the project, the defects date or may even include the 6, 10 or 12-year limitation period following completion of the works to cover any liability for potential latent defects, the expiry date being defined within the Scope.

Parent company guarantees are normally used as an alternative to a performance bond (Option X13).

Such a guarantee is cheaper than a performance bond, as the Contractor will normally just charge an administration fee rather than in the case of a performance bond, where the Contractor is actually paying a premium for an insurance policy, but it may give less certainty of redress because it is not supplied by an independent third party so it is dependent on the survival, and the ability to pay, of the parent company.

However, whilst accepting less independence, parent company guarantees for the proper performance of the contract can be more advantageous than bonds. Rather than receiving a fixed amount in compensation, the parent company is normally obliged to either complete the works in accordance with the Contractor's original obligations on behalf of its subsidiary company, or fund the completion of the contract by others, so effectively completion is guaranteed.

In addition, further recompense can be sought for time delays in completion through the normal clauses incorporated in the contract. In the case of performance bonds the "guarantee" is that a sum of money is available to at least partly compensate the Client in the event of the Contractor's default.

Because the financial strength of the parent company may be linked to that of the Contractor, a parent company guarantee will be acceptable only if the parent company (or holding company) is financially strong and its financial resources are largely independent of those of the Contractor. Obviously, if the insolvency is not limited to the Contractor but also related to the parent group a parent company guarantee will be virtually useless, save for any entitlement under insolvency law.

Option X5: Sectional Completion

If the Client requires sections of the works to be completed by the Contractor before the whole of the works are completed, then Option X5 should be chosen.

References in the contract to the works, Completion and Completion Date will then apply to either the whole of the works or a defined section in Contract Data Part 1, which should give a description of each section and the date by which it is to be completed.

Option X5 may be selected together with Option X6 (Bonus for Early Completion) and/or Option X7 (Delay Damages).

Option X6: Bonus for early Completion

Often early completion would be of benefit to the Client, for example the Contractor is building a shopping mall and early completion would allow tenants to move in earlier which would provide early revenue for the Client.

Option X6 provides for the Client to pay a bonus to the Contractor for early Completion.

The amount of the bonus is stated by the Client on a "per day" basis in Contract Data Part 1 and is calculated from the earlier of Completion and when the Client takes over the works, until the Completion Date.

Option X7: Delay damages

Delay damages in the ECC are normally referred to in other contracts as "liquidated damages".

Delay damages are pre-defined amounts inserted into Contract Data Part 1 and paid or withheld from the Contractor in the event that it fails to complete the works by the Completion Date.

Within most legislation, the amount included within the contract for delay damages should be a "genuine pre-estimate of likely losses" – not a penalty.

Many contracts require the contract administrator (Engineer, Architect, etc) to issue some form of certificate confirming that the Contractor failed to complete on time, and also for the Client to notify the Contractor in writing that it will be withholding the relevant damages, but the ECC contract provides for the Contractor to pay delay damages at the rate stated in the Contract Data until Completion or the date on which the Client has taken over the works, whichever is earlier, without having to certify that the Contractor has defaulted.

Also, many contracts require the contract administrator to value the works and the Client then deducts the liquidated damages from the amount due to the Contractor, but the ECC contract requires the Project Manager to deduct amounts to be paid or retained from the Contractor within its assessment and certificate.

Under Clause X7.3, if the Client takes over the works before Completion, the Project Manager assesses the benefit to the Client of taking over that part of the works as a proportion of taking over the whole of the works and the delay damages are reduced in this proportion.

Whilst it is probably the most correct way to assess remaining delay damages, the Project Manager having to assess the benefit to the Client can at best be a subjective exercise and may possibly lead to disputes with the Contractor.

Most other contracts state that the amount of liquidated damages is reduced by the same proportion that the part that has been taken over bears to the value of the whole of the works, so if a third of the value of the works has been taken over, the amount of the liquidated damages is reduced by a third.

Option X8: Undertakings to the Client or Others

This is a new Option, not formerly included within the NEC3 ECC, but in the NEC3 Professional Services Contract as "Collateral Warranty Agreements", and provides for the Contractor to give undertakings to Others as stated in the Contract Data and, if required, in the form stated in the Scope.

This may also include undertakings between a Subcontractor and Others if required by the Contractor. Typically such documents are often referred to as collateral warranties.

The Client prepares the undertakings and sends them to the Contractor for signature, the Contractor signs or arranges for the Subcontractor to sign them, and returns them to the Client within three weeks.

Whilst stating that Option W8 was previously included in the NEC3 Professional Services Contract as "Collateral Warranty Agreements", the Option within the ECC does not specifically refer to collateral warranty agreements. If parties wish to enter into collateral warranty agreements, then they need to address that matter separately, with appropriate documentation to suit the parties (see also Chapter 2).

Option X9: Transfer of rights

This is a new Option, again as with Option X8, not formerly included within the NEC3 ECC, but in the NEC3 Professional Services Contract. The Client owns the Contractor's rights over materials prepared for the design of the works except as stated otherwise in the Scope.

The Contractor obtains other rights for the Client as stated in the Scope and also obtains from Subcontractors equivalent rights for the Client.

Option X10: Information modelling

This is a new Option, not formerly included within the NEC3 contract, which provides for Building Information Modelling (BIM).

Some defined terms should be considered in reviewing Option X10.

- The Information Plan is submitted by the Contractor to the Project Manager who then accepts or does not accept within two weeks of receiving it. The Information Execution Plan is then defined as the *information execution plan* (identified within the Contract Data), or the latest submitted by the Contractor and accepted by the Project Manager.
- Project Information is provided by the Contractor and used to create or change the Information Model.
- The Information Model is the electronic integration of the Project Information, and other information provided by the Client and other Information Providers. The Client is liable for any fault or error in the Information Model, unless there is a Defect in the Project Information provided by the Contractor. The Contractor is required to provide insurance against claims made in respect of failure to provide the Project Information using reasonable skill and care.
- The Information Model Requirements are the requirements identified in the Scope for creating or changing the Information Model.
- Information Providers are the people or organisations who contribute to the Information Model.

There is also provision within the Option for early warnings where something could affect the Information Model, and for the Contractor to include within its quotation for a compensation event where the Information Execution Plan is altered by a compensation event.

Option X11: Termination by the Client

This Option, not explicitly included within NEC3, though there was some provision for it, provides for the Client to terminate the Contractor's employment for a reason not stated in the Termination Table, by notifying the Project Manager and Contractor.

If the Client does terminate, it may complete the service and use any Plant and Materials provide by the Contractor (Procedure P1), and the Contractor leaves the Service Areas and removes the Equipment (P4).

In respect of payments, the Contractor is entitled to payments as if it had terminated:

Amount A1

- an amount due as for normal payments.
- the Defined Cost for Plant and Materials which have been delivered and retained by the Client or which the Client owns and for which the Contractor has to accept delivery.
- other Defined Cost in reasonable expectation of completing the whole of the service (such as long-term supply contracts for consumables).
- any amounts retained by the Client.

Amount A2

- the forecast cost of removing Equipment.

Amount A4

- for Options A and C, the fee percentage applied to the difference between the original total of the Prices and the Price for Service Provided to Date.
- for Option E, the fee percentage applied to the difference between the first forecast of the Defined Cost for the service and the original total of the Prices and the Price for Service Provided to Date less the Fee.

Option X12: Multiparty collaboration (not used with Option X20)

This Option, formerly called "*Partnering*" in NEC3, enables a multiparty partnering agreement to be implemented.

In this case Option X12 is used as a Secondary Option common to the contract which each party has with the body which is paying for the work.

The content of Option X12 is derived from the "Guide to Project Team Partnering" published by the Construction Industry Council (CIC). It is estimated that Option X12 is used on less than 10% of NEC3 contracts.

It must be stressed that no legal entity is created between the Partners, so it is not a partnership as such.

Some definitions need to be explained:

(i) The Partners are those named in the Schedule of Partners.
(ii) An Own Contract is a contract between two Partners.
(iii) The Core Group comprises the Partners selected to take decisions on behalf of the Partners.
(iv) The Schedule of Core Group Members is a list of Partners forming the Core Group.
(v) Partnering Information is information which specifies how the Partners work together.
(vi) A Key Performance Indicator is an aspect of performance for which a target is attached in the Schedule of Partners.

Each Partner, represented by a single representative, is required to work with the other Partners in accordance with the Partnering Information to achieve the Promoter's (normally their joint Client) objective stated in the Contract Data and the objectives of every other Partner.

The Core Group acts and takes decisions on behalf of the Partners. The Core Group also keeps up to date a Schedule of Core Group Members and a Schedule of Partners.

The Partners are required to work together, using common information systems, and a Partner may ask another Partner to provide information which it needs to carry out the work in its Own Contract.

The Core Group may give an instruction to the Partners to change the Partnering Information, which is a compensation event.

The Core Group also maintain a timetable showing the Partners' contributions. If the Contractor needs to change its programme to comply with the timetable then it is a compensation event.

Each Partner also gives advice, information and opinion to the Core Group where required.

Each Partner must also notify the Core Group before subcontracting any work, though it does not say that the Core Group is required to respond to the notification.

Finally Option X12 provides for Key Performance Indicators (KPI) with amounts paid as stated in the Schedule of Partners. The Promoter may add a KPI to the Schedule of Partners but cannot delete or reduce a payment.

Option X13: Performance bond

A performance bond is an arrangement whereby the performance of a contracting party (the Principal) is backed by a third party (the Surety), which could be a bank, insurance company or other financial institution, that should the Principal fail in its obligations under the contract, normally due to the Principal's insolvency, the Surety will pay a pre-agreed sum of money to the other contracting party (the Beneficiary). Whilst bonds are not always linked to insolvency, contracts normally give remedies in the event of default, e.g. defects provisions, retention, withholding of payment, delay damages, etc. It is where the defaulter is not able to provide a remedy because it is insolvent that bonds normally provide the remedy.

This bond may be between any two contracting parties, Client and Contractor, Client and Consultant, Contractor and Subcontractor, or any other contractual relationship.

Note that as these are conditional bonds, it is normally a condition that default has to have occurred, there is no other means of remedy or recompense and the amount claimed is directly related to the default, before payment can be made to the beneficiary.

Clearly, if a bank provides a bond there may be some implications with regard to further credit as a safeguard should the bond be called, with the bank also pursuing any amount paid from their customer, whereas with an insurance company the risk is accepted through the policy.

If a performance bond is required from the Contractor it should be provided by the Contract Date or within four weeks of the Contract Date.

The amount of the bond, usually 10% of the contract value must be stated in Contract Data Part 1 and the form of the bond must be in the form stated in the Scope. In the event of the default by the Contractor the Client is not paid the full amount of the bond, but the amount of the bond is the maximum amount available to meet the Client's costs incurred as a direct result of the default.

Normally, the value of the performance bond does not reduce, but they should have an expiry date, which could be completion of the project, the Defects Date or may even include the 6, 10 or 12-year limitation period following completion of the works to cover any liability for potential latent defects, the expiry date being defined within the Scope.

It is worth mentioning that payment of a performance bond is often very difficult and subject to a number of conditions, for example, clear evidence of the Contractor's inability to perform its obligations, and also unequivocal evidence of the Client's loss as a direct result of the Contractor's default, which in truth may not be fully known for some time after the Contractor's insolvency. Clearly the expiry date for the bond has to be kept in mind if it takes some time to ascertain the Client's actual loss, as the right to claim may have expired.

A performance bond may be provided by a bank, an insurance company or a specialist bond provider. Clearly, if a bank provides a bond there may be some implications with regard to further credit as a safeguard should the bond be called, with the bank also pursuing any amount paid from their customer, whereas with an insurance company the risk is accepted through the policy.

The bank or insurer which provides the performance bond must be accepted by the Project Manager.

It is important to note the difference between an insurance policy and a bond, in that an insurance policy is a contract between TWO parties, the insured and the insurer, the insurer guaranteeing that it will pay a third party, who may not be specifically named, should an event occur, whilst a bond is a contract between THREE parties, the Contractor, the Client and the Surety, all of whom are named within the agreement, with no other party being able to benefit from the bond.

A performance bond will not of itself ensure that contracts are carried out efficiently and to time, but it will be one of the number of commercial pressures on the Contractor to perform well. A performance bond can provide some compensation if the Contractor defaults on its obligations.

The cost of the performance bond will normally be governed by:

(i) The technical ability of the Contractor.
(ii) The usual type of work undertaken by the Contractor.
(iii) The size of the project.
(iv) Contracts already bonded for the Contractor.
(v) The overall management of the Contractor's business.
(vi) The financial stability of the Contractor.

There are other types of bond available:

(i) Advanced payment bond
    This covers the Contractor's repayment of an advanced payment from the Client to the Contractor. This is covered in the contract by Option X14.
(ii) Payment bond
    This covers the Client's duty to pay the Contractor, or the Contractor's duty to pay its Subcontractors.

(iii) Bid or tender bond

Very often on particularly large projects with a long lead-in time, a bond may be required to ensure that the successful tenderer will not withdraw its tender and is able to proceed when required, in terms of its other commitments and its solvency.

In a similar vein, a scenario may be where a Contractor prices a tender based on a Subcontractor's price and finds upon acceptance of its tender that the Subcontractor's offer is no longer open for acceptance, and the Contractor may find itself having to place orders with another Subcontractor for a higher price.

It may be worthwhile in this event to bond the Subcontractor so that in the event that the acceptance period expires, the Contractor can recover monies through the bond.

(iv) Retention Bonds

Again, on particularly large projects it may be more practical to release retention monies early with a bond in place to protect the parties in the event of defects arising.

UNCONDITIONAL OR ON DEMAND BONDS

"Unconditional" or "on demand bonds" can be redeemed by the beneficiary, whether or not there has been a default within the contract and whether or not there has actually been a loss at all!

Most Clients will refrain from calling in bonds unnecessarily, but the entitlement remains should they wish to.

The "knock on" effect of calling in these bonds without due cause is that Contractors will price for that possibility within their tenders, as someone has to pay, and the result will be escalating prices. Insurers are naturally very reluctant to accept on demand bonds.

CREATION OF BONDS

Bonds have to be made in writing and will be in the form of a deed, signed and sealed by the contracting parties.

The duration and financial limits of the Bond MUST be clarified and stated within the deed.

RELEASE AND CANCELLATION OF BONDS

The deed executed will normally specify the course of action to be taken in the event of failure of the Principal leading to the calling in of the Bond by the Beneficiary.

Sometimes the Beneficiary has to give a form of notice to the Principal or it may have to sue for non-performance.

Only the Beneficiary can cancel the Bond, the Principal issuing a request to him, once it has "performed".

Option X14: Advanced payment to the Contractor

This Option is appropriate when the Contractor will incur significant "up front" costs before it starts receiving payments, for example in pre-ordering specialist materials, plant or equipment.

The payment is made within the first payment assessment or if an advanced payment bond is required, at the next assessment after the Client receives the bond.

If an advanced payment bond is required it is issued by a bank or insurer which the Project Manager has accepted, the bond being in the amount that the Contractor has not repaid.

Advance payment bonds are a helpful security when an advance payment is made to a Contractor for works to be performed; the Project Manager must accept the provider of that bond.

The amount of the repayment instalments is stated in Contract Data Part 1.

Option X15: The Contractor's design

This is an Option, formerly in NEC3 called "*Limitation of the Contractor's liability for his design to reasonable skill and care*". There are several items covered by this Option.

Firstly, whilst the Scope defines what, if any design, is to be carried out by the Contractor, the contract is silent on the standard of care to be exercised by the Contractor when carrying out any design.

Two terms that relate to design liability are "fitness for purpose" and "reasonable skill and care".

Whilst this book is intended for international use, in defining the term "fitness for purpose" one must look in English law to the Sale of Goods Act 1979, which refers to the quality of goods supplied including their state and condition complying in terms of "fitness for all the purposes for which goods of the kind in question are commonly supplied".

In construction, fitness for purpose means producing a finished project fit in all respects for its intended purpose. This is an absolute duty independent of negligence, and in the absence of any express terms within the contract to the contrary, a Contractor who has a design responsibility will be required to design and build the project "fit for purpose".

Some contracts will limit the Contractor's liability to that of a consultant, i.e. reasonable skill and care.

The ECC does this through Option X15; if this Secondary Option is not chosen, the Contractor's liability for design is "fitness for purpose". If the Contractor corrects a Defect for which it is not liable, it is a compensation event.

Contrast this with the level of liability of a consultant (whether acting for a Client or a Contractor) in providing a design service, which is defined, again in English law by the Supply of Goods and Services Act 1982, where there is an implied term that the consultants will carry out the service, in this case design, with reasonable care and skill, which means designing to the level of an ordinary, but competent person exercising a particular skill.

So, in the absence of any express terms to the contrary, a designer will normally be required to design using "reasonable skill and care". This is normally achieved by the designer following accepted practice and complying with national standards, codes of practice, etc.

Clearly, despite the Contractor believing that it has offset its design obligations and liability to its designing consultant has to be aware that it, and its consultant, have differing levels of care and liability.

The remaining items covered by this Option are:

- The Contractor may use material provided to it under the contract, unless the ownership of the material has been given to the Client.
- The Contractor retains copies of drawings, specifications, reports, etc in the form stated in the Scope for the period of retention, normally 6 or 12 years after Completion of the works.
- The Contractor may be required to provide Professional Indemnity (PI) Insurance if required, and in the amount in the Contract Data.

Option X16: Retention (not used with Option F)

The Client may retain a proportion of the Price for Work Done to Date once it has reached any retention free amount, the retention percentage and any retention free amount being stated in Contract Data Part 1.

Defined Cost does not include amounts deducted for retention, so, if the Contractor deducts retention from a Subcontractor, the figure before the deduction is used in calculating Defined Cost, ensuring that there is no "double deduction" of retention.

Following Completion of the whole of the works, or the date the Client takes over the whole of the works, whichever happens earlier, the retention percentage is halved, and then the final release is upon the issue of the Defects Certificate.

It is important to note that retention is held against undiscovered defects and not incomplete work.

If stated in the Contract Data, or if agreed by the Client, the Contractor provides a retention band provided by a bank or insurer, accepted by the Project Manager, and in the form stated in the Scope.

Option X17: Low performance damages

In the event that the Contractor produces defective work, the Client has three options:

 (i) The Contractor corrects the Defect (Clause 44.1).
 (ii) If the Contractor does not correct the Defect, the Project Manager assesses the cost to the Client of having the Defect corrected by other people and the Contractor pays this amount (Clause 46.1).
(iii) The Client can accept the Defect and a quotation from the Contractor for reduced Prices and/or an earlier Completion Date (Clause 45).

Where the performance in use fails to reach the specified level within the contract, the Client can take action against the Contractor to recover any damages suffered as a result of the breach, but as an alternative can recover low performance damages under Option X17 if it has been selected.

Example

The Scope requires the Contractor to design and install a HVAC system to a major retail development.

There are specific and measurable performance criteria for the system including temperature variations and energy efficiency. The system is required to have a design life of 30 years (360 months).

The Scope states that the system will be tested at Completion and includes a table showing how the performance of the system will be measured and acceptable levels of achievement.

The system is expected to perform to 98 – 100% of performance criteria. If it falls within 90 – 98% the system will be accepted, but low performance damages will be payable by the Contractor to the Client. If it falls below 90% it will not be accepted.

Contract Data Part 1 contains the following entries:

amount per month	performance level
£50.00	96% – 98%
£150.00	94% – 96%
£200.00	92% – 94%
£250.00	90% – 92%

When tested, the system achieves 95% of the performance criteria.

Therefore, 360 months × £150.00 = £54,000 is payable by the Contractor at the Defects Date.

Option X18: Limitation of liability

Under this Option, a limit may be placed on the Contractor's liability for the following:

- liability for the Client's indirect or consequential loss.
- liability for causing any loss of or damage to the Client's property.
- liability to the Client for latent defects due to its design.
- total liability for all matters under the contract, other than excluded matters in contract, tort or delict.

Excluded matters are:

- loss or damage to Client's property.
- delay damages if Option X7 applies.
- low performance damages if Option X17 applies.
- Contractor's share if Option C or D applies.

The Contractor is not liable for any matter unless it has been notified to the Contractor before the end of liability date which is stated in the Contract Data in terms of years after the Completion of the whole of the works.

In the UK, this may be 6 or 12 years, dependent on the type of contract; other legislations set this at 10 years.

Option X20: Key Performance Indicators

Performance of the Contractor can be monitored and measured against Key Performance Indicators (KPIs) using Option X20.

Targets may be stated for Key Performance Indicators in the Incentive Schedule.

From the starting date until the Defects Certificate is issued, the Contractor is required to report its performance against KPIs to the Project Manager at intervals stated in the Contract Data including the forecast final measurement. If the forecast final measurement will not achieve the target stated in the Incentive Schedule the Contractor is required to submit its proposals to the Project Manager for improving performance.

The Contractor is paid the amount stated in the Incentive Schedule if the target for a KPI is improved upon or achieved. Note that there is no payment due from the Contractor if it fails to achieve a stated target.

The Client may add a new KPI and associated payment to the Incentive Schedule but may not delete or reduce a payment.

Option X21: Whole life cost

This is a new Option, not formerly included within the NEC3 contract.

Under this Option, the Contractor may propose to the Project Manager that the Scope is changed in order to reduce the cost of operating and maintaining the asset.

If the Project Manager is prepared to consider the change, the Contractor submits a quotation to the Project Manager which includes:

- a detailed description,
- the forecast cost reduction to the Client of the asset over its whole life,
- an analysis of the resulting risks to the Client,
- the proposed change to the Prices, and
- a revised programme showing any changes to the Completion Date and Key Dates.

The Project Manager consults with the Contractor about the quotation and replies within the period for reply either accepting or not accepting the quotation.

If the quotation is accepted, the Project Manager changes the Scope, the Prices, the Completion Date and any Key Dates and accepts the revised programme. The change to the Scope is not a compensation event.

Option X22: Early Contractor involvement (used only with Options C and E)

This is a new Option, not formerly included within the NEC3 contract.

Option X22 commences with four identified and defined terms:

1 Budget – the items and amounts stated in the Contract Data.
2 Project Cost – the total paid by the Client for the items stated in the Budget.
3 Stages One and Two – as stated in the Scope.
4 Pricing Information – the information which specifies how the Contractor prepares its assessment of the Prices for Stage Two.

The Contractor is required to provide detailed forecasts of the total Defined Cost of the work to be done in Stage One for acceptance by the Project Manager. These forecasts are prepared on a periodic basis commencing from the starting date at intervals stated in the Contract Data.

The Project Manager has one week in which to accept/not accept each forecast.

The Contractor is required, in consultation with the Project Manager, to provide forecasts of the Project Cost and submit them to the Project Manager. These forecasts are prepared on a periodic basis commencing from the starting date until completion of the whole of the works at intervals stated in the Contract Data.

The Contractor submits its design proposals for Stage Two, including a forecast of the effect of the design proposals on the Project Cost and the Accepted Programme, to the Project Manager for acceptance as stated within the submission procedure within the Scope. If the Project Manager does not accept he gives reasons and the Contractor resubmits.

The Project Manager issues a notice to proceed to Stage Two when the Contractor has obtained approvals and consents from Others, changes to the Budget have been agreed, the Project Manager and Contractor have agreed the total of the Prices for Stage Two, and the Client has confirmed that the works are to proceed.

If the Project Manager does not issue a notice to proceed to Stage Two, the Client may appoint another Contractor to complete Stage Two.

If the Project Manager issues an instruction changing the Client's requirements, the Project Manager and the Contractor agree changes to the Budget within four weeks.

A budget incentive is paid to the Contractor if the final Project Cost is less than the Budget.

Option Y(UK)1: Project Bank Account

In 2008, the Office of Government Commerce (OGC) published a guide to fair payment practices, following which the NEC Panel prepared a document in June 2008 to allow users to implement these fair payment practices into NEC contracts.

This is now Option Y(UK)1, which provides for a Project Bank Account to be set up which receives payments from the Client which in turn is used to make payments to the Contractor and Named Suppliers. The Project Bank Account is established within three weeks of the Contract Date.

There is also a Trust Deed between the Client, the Contractor and Named Suppliers containing the necessary provisions for administering the Project Bank Account.

The Contractor should include in any subcontracts for Named Suppliers to become party to the Project Bank Account through a Trust Deed via a Joining Deed.

The Contractor notifies the Named Suppliers of the details of the Project Bank Account and the arrangements for payment of amounts due under their contracts.

The Named Suppliers will be named within the Contractor's tender, but also additional Named Suppliers may be included subject to the Client's acceptance by means of a Joining Deed, which is executed by the Client, the Contractor and the new Named Supplier. The new Named Supplier then becomes a party to the Trust Deed.

As the Project Bank Account is maintained by the Contractor, he pays any bank charges and also is entitled to any interest earned on the account. The Contractor is also required at tender stage to put forward his proposals for a suitable bank or other entity which can offer the arrangements required under the contract.

The process every month is that the Contractor submits an application for payment including details of amounts due to Named Suppliers in accordance with their contracts.

The Client makes payment to the Project Bank Account, the Contractor makes payment to the Project Bank Account of any amounts which the Client has notified the Contractor intends to withhold from the certified amount and which is required to make payment to Named Suppliers.

The Contractor then prepares the Authorisation, setting out the sums due to Named Suppliers. After signing the Authorisation, the Contractor submits it to the Client for signature and submission to the project bank.

The Contractor and Named Suppliers then receive payment from the Project Bank Account of the sums set out in the Authorisation after the Project Bank Account receives payment.

In the event of termination, no further payments are made into the Project Bank Account.

Option Y(UK)2: The Housing Grants, Construction and Regeneration Act 1996

This option is only applicable to UK contracts where the Housing Grants, Construction and Regeneration Act 1996 (as amended by the Local Democracy, Economic Development and Construction Act 2009) applies, and deals only with the payment aspects of the Act, adjudication under the Act being covered by Option W2.

It is important to note that as the Local Democracy, Economic Development and Construction Act 2009 amends the Housing Grants, Construction and Regeneration Act 1996, you have to take into account both Acts to fully understand how it applies to your contract, also that any Z clauses must comply, where applicable, and particularly with respect to payment, must comply with the Act.

Although it is an "Option", where the Act applies, it must be selected, failing which the Parties are required to comply with the law anyway.

The main features of Option Y(UK)2 are as follows (time periods are mostly stated in days, note that time periods within the core clauses and other Main Options are mostly stated in weeks):

DATES FOR PAYMENT

The due date for payment is seven days after the assessment date.

The date on which the final payment becomes due is:

- if the Project Manager makes an assessment after the issue of a Defects Certificate – five weeks after its issue.
- if the Project Manager does not make an assessment after the issue of the Defects Certificate – one week after the Contractor issues its assessment.
- if the Project Manager has issued a termination certificate – fourteen weeks after its issue.

The final date for payment is fourteen days after the payment becomes due, unless there is a different period stated in the Contract Data.

The Project Manager's certificate is the notice of payment (which may be "zero"). If the Project Manager does not make an assessment after the issue of a Defects Certificate the Contractor's assessment is the notice of payment.

PAY LESS NOTIFICATION

If either Party intends to pay less than the notified sum, it notifies the other Party not later than seven days before the final date of payment. This is a pre-condition to paying less than the notified sum.

If the Client terminates for reasons R1 to R15, R18 or R22, and a certified payment has not been made at the date of the termination certificate, the Client makes the payment unless:

- a pay less notice has been issued, or
- the termination is for reasons R1 to R10, and the reason occurred after the last date on which it could have notified the Contractor of a pay less notice.

SUSPENSION OF PERFORMANCE

If the Contractor exercises its right under the Act to suspend performance it is a compensation event.

Option Y(UK)3: Contracts (Rights of Third Parties) Act 1999

This option is only applicable to UK contracts where the Contracts (Rights of Third Parties) Act 1999 applies.

The principle of privity of contract is that a person who is not a party to a contract cannot enforce the terms of that contract.

The Contracts (Rights of Third Parties) Act 1999 effectively abolishes privity of contract as an absolute principle, enabling a person who is not a party to a contract, a "third party", to enforce a term of the contract if the contract expressly provides that it may. The third party must be expressly identified in the contract, either by name or as a member of a class or by a particular description, but does not need to be in existence at the date of the contract.

Under the Act, there is available to the third party any remedy that would have been available to him in an action for breach of contract as if it had been a party to the contract.

The parties to the contract are prohibited from making any agreement to vary the contract in such a way as to adversely affect the third party's entitlement to enforce the contract without the third party's consent.

The Act names two specific parties:

- "the promisor" means the party to the contract against whom the term is enforceable by the third party, and
- "the promisee" means the party to the contract by whom the term is enforceable against the promisor.

While the Act is not specially aimed at the construction industry, it is an industry which spends a lot of time and money in constructing collateral warranties enabling a multitude of parties to have certain rights over specified periods of time, so it is now possible to draft building contract documents to confer benefits upon the parties who would have had the benefit of collateral warranties.

The construction industry has generally not been in favour of this act, preferring to use collateral warranties, but it must be said that the Act has several advantages in terms of time and cost in that only one document needs to be drafted, though it is fair to say that many parties prefer a warranty as specific evidence of their rights. Banks and funders in particular like the comfort of a good old-fashioned document.

Option Z: Additional conditions of contract

Option Z allows conditions to be added to or omitted from the core clauses.

All changes to the core clauses should be included as Z clauses rather than amending the core clauses themselves, so in effect the clause remains in the contract, but is amended within the Z clause.

It is also critical that when drafting a Z clause, it must be clearly stated what happens to the original core clause, for example, if it is deleted. If the original core clause is not deleted, then it is likely that an inconsistency will arise which will be interpreted against the party who wrote or amended the clause, i.e. the Client.

These conditions may modify or add to the core clause, to suit any risk allocation or other special requirements of the particular contract. However, changes should be kept to a minimum, consistent with the objective of using industry standard, impartially written contracts.

It must be remembered that if a Client amends a contract to allocate a risk to the Contractor which may have been intended to be held by the Client, the Contractor has a right to price it in terms of time and money, therefore the practice of Clients amending contracts to pass risk to Contractors without considering who is best able to price, control and manage those risks can in many cases prove to be unwise and uneconomical.

It is important to spend time considering whether a Z clause is appropriate in each case, then when that decision has been made, that the clause is drafted correctly and aligned to the drafting principles of the original contract, in the case of the ECC using ordinary language, present tense, correct capitalisation, correct italics, short sentences and bullet pointing.

It is critical that whoever drafts Z clauses is fully aware of the NEC drafting philosophy (see section above, Clarity and simplicity). It is not unusual to see Z clauses in an ECC contract written in a legalistic language, in the future tense and without punctuation, apart from full stops!

Some examples of Z clauses that have been used with the NEC3 Engineering and Construction Contract.

N.B. The author is not recommending the use of these Z clauses, but simply quoting them as examples.

2 The Contractor's main responsibilities

Subcontracting 26

Reason for Z clause: Many Project Managers would like to give a reason for not accepting a proposed Subcontractor, that his appointment "may not" rather than "will not" allow the Contractor to Provide the Works, to allow some subjectivity. Therefore it is sometimes amended:

Delete:

Second sentence of Clause 26.2

Insert:

A reason for not accepting the Subcontractor is that his appointment may not allow the Contractor to Provide the Works.

4 Testing and Defects

Searching for and notifying Defects

Reason for Z clause: Why does the Supervisor need to know the Contractor has found a Defect?

Clause 42.2

Delete:

The words "and the Contractor notifies the Supervisor of each Defect as soon as he finds it".

5 Payment

The Contractor's share

Reason for Z clause: Most people (including me) believe that the Contractor's share should be calculated when the Price for Work Done to Date reaches the current Total of the Prices (the target) rather than on Completion, so that the Contractor is not paid monies that he will have to pay back later. For that reason, users tend to amend the clause as follows:

Delete:

Clause 53.3

Insert:

The *Project Manager* makes a preliminary assessment of the *Contractor's* share at any assessment date. If the forecast Price for Work Done to Date is

(continued)

(continued)

less than the final total of the Prices, this share is included in the amount due following completion of the whole of the works (you can leave this sentence in if you wish to give it balance so that the Contractor has the right to early payment of a "gain" share as well as a "pain share"). If the forecast Price for Work Done to Date is more than the forecast total of the Prices, this share is deducted from the amount due.

For Options C, D, E and F only

Reason for Z clause: Many Clients and Project Managers do not accept Clause 11.2(29), in that a forecast has to be made of the Defined Cost up to the next assessment date. For that reason, users tend to amend the clause to the previous clause under NEC2, as follows:

Delete Clause 11.2(29)

Insert:

The Price for Work Done to Date is the total Defined Cost which the *Contractor* has paid plus the Fee.

Note NEC2 used the words "Actual Cost" rather than "Defined Cost".

For Options C, D and E

Reason for Z clause: Many Clients and Project Managers do not accept that if a Contractor corrects a Defect before Completion it is not a Disallowed Cost, therefore it is payable. If this is not acceptable then suggest:

Clause 11.2(25) 6th Bullet

Delete:

the words "after Completion"

Additional prompt payment clause

Prompt payment to Subcontractors

Where the Contractor enters into a subcontract with a Supplier or Subcontractor he includes a term to be included in the subcontract which requires payment to be made to the Supplier or Subcontractor within a specified period not exceeding thirty days from receipt of a valid invoice as defined by the subcontract requirements.

6 Compensation events

Assessing compensation events 63

Reason for Z clause: The default position with pricing compensation events is that they are assessed as the effect of the compensation event upon actual/forecast Defined Cost plus the resulting Fee (Clause 63.1), *but* if the Project Manager and Contractor agree, rates and lump sum prices may be used instead.

These clauses are sometimes reversed so that compensation events are based on agreed rates and prices, e.g. bill rates, etc, *unless* the Project Manager and Contractor agree to use Defined Cost plus the resulting Fee.

8 Risks and insurance

Insurance cover

Reason for Z clause: Professional Indemnity (PI) insurance is not specifically required under the contract, but especially in a design and build contract, this may be required. Note that you would have to include reference to the amount of Professional Indemnity Insurance required within the Contract Data.

Insert:

The *Contractor* maintains Professional Indemnity (PI) insurance for the amount stated in the Contract Data for each and every claim, and with exclusions and excesses in line with those normally applicable to contractors undertaking the design of works such as the *works*, for 12 years following Completion of the whole of the *works*, provided that such insurance continues to be available generally to contractors such as the Contractor on reasonable commercial rates.

0.3 The roles of the parties

There are a number of parties referred to in the contract:

- the Client
- the Contractor
- the Project Manager
- the Supervisor
- the Subcontractor
- the Adjudicator

The Client

Although the contract is between the Client and the Contractor, the Client is rarely mentioned in the contract.

The Client's obligations are:

- to act as stated in the contract and in a spirit of mutual trust and co-operation (Clause 10.2).
- to allow the Contractor access to and use of each part of the Site for the Contractor to carry out the work included in the contract. This access must be allowed on or before the later of the access date (the date in the Contract Data) and the date for access shown on the Accepted Programme (Clause 33.1).
- to allow the Contractor access to and use of a part of the works which it has taken over if they are needed for correcting a Defect (Clause 46.1).
- to make payment to the Contractor in accordance with the Project Manager's certificates (Clause 51.2).
- to take over the works at Completion (Clause 35.1).

The Contractor

The Contractor and the Client are the Parties to the contract, the Contractor's responsibility being to Provide the Works in accordance with the contract, the Scope describing the works and any constraints on how it has to provide it.

There are also obligations in respect of complying with instructions, notifying early warnings, submitting programmes for acceptance, and notifying and submitting quotations for compensation events which are highlighted in later chapters.

The Project Manager

The Project Manager is appointed by the Client and manages the contract on its behalf. Other forms of contract tend to name an Engineer or an Architect, who acts for the Client, and in addition to having design responsibilities, also administers the contract. The Engineering & Construction Contract separates the role of design from that of managing the project and administering the contract.

The Project Manager is named in Contract Data Part 1 and may be appointed from the Client's own staff or may be an external consultant. The Project Manager is a named individual, not a company.

It is vital that the Client appoints a Project Manager who has the necessary knowledge, skills and experience to carry out the role, which includes acceptance of designs and programmes, certifying payments and dealing with compensation events.

Disciplines which have carried out the Project Manager role to date include Engineers from a civil engineering, structural or process background, Architects, Building Surveyors and Quantity Surveyors in addition to Project Managers themselves.

It is also vital that the Client gives the Project Manager full authority to act for him, particularly when considering the timescales imposed by the contract. If the Project Manager has to seek approvals and consents from the Client, then it must do so, and the appropriate authority must be given, in compliance with the contractual timescales. If the approval process is likely to be lengthy, it is critical that both the Client and its Project Manager set up an accelerated process to comply with the contract or that the timescales in the contract are amended. The former is the vastly preferred method in order that NEC4 will work to its full effect.

Although the Project Manager manages the contract at the post-contract stage, it is usually appointed pre-contract to deal with matters such as feasibility issues, advising on design, procurement, cost planning, tendering and programme matters. Note that, unlike many other contracts, the ECC does not name the Quantity Surveyor, so this would need to be considered in dealing with financial matters pre- and post-contract. Either the Project Manager may carry out the role of the Quantity Surveyor itself or delegate to another.

The role of a Project Manager within the ECC is different to the usual concept of a project manager and it is important that this difference is recognised as it is a traditional area of misunderstanding amongst users of the contract.

There should only be one Project Manager for the project, though the Project Manager can delegate responsibilities to others after notifying the Contractor and may subsequently cancel that delegation (Clause 14.2). One assumes that the Project Manager will also notify the Client, if he expects the Client to pay for the delegate!

As with any other notification under the contract, if the Project Manager (or the Supervisor) wishes to delegate, the notice to the Contractor must be in a form which can be read, copied and recorded, i.e. by letter or email, not by verbal communication such as a telephone call.

The delegation may be due to the Project Manager being absent for a period due to holidays or illness, or because it wishes to appoint someone to assist him in its duties.

Although the contract is not specific, the notice should identify who the Project Manager is delegating to, what their authority is, and how long the delegation will last, so the Contractor is in no doubt as to who has authority under the contract.

It is important to recognise that in delegating, the Project Manager is sharing an authority with the delegate who will then represent him; it is not passing responsibility on to them. The authority is shared, and the Project Manager may still take an action which they have delegated, but ultimately responsibility will remain with the Project Manager as the principal.

It is also important to note that one can only delegate outwards from the principal, i.e. no-one can assume authority, possibly because, within an organisation, they are senior to the Project Manager.

If the Client wishes to replace the Project Manager or the Supervisor it must firstly notify the Contractor of the name of the replacement before doing so.

Designers

Designers are not named parties in the ECC as the design may be carried out on behalf of the Client and/or the Contractor. The Client is required to appoint designers for its own design, and the Contractor for its design. The concept of novation of designers is considered in Chapter 2. It is important to recognise that the designer does not have the same authority as, say, an Architect under the JCT contracts who has the dual role of Designer and Contract Administrator. Under the ECC the Designer only designs, it does not administer the contract, and therefore any issuing of drawings, other design information and associated instructions must be done through the Project Manager.

The Supervisor

The Supervisor is appointed by the Client and manages issues regarding testing and defects on its behalf. Again, as with the Project Manager, it is named in Contract Data Part 1 and may be appointed from the Client's own staff or may be an external consultant. Again, as with the Project Manager, the Supervisor is a named individual not a company. It is possible that, provided it has sufficient expertise and time available, the Supervisor and the Project Manager could be the same person. It is not unusual for the Client's Designer and the Supervisor to be the same person as it is well qualified, having done the design, to inspect the works to make sure the Contractor has complied with it.

The title Supervisor is not found in other contracts, many seeing the role as similar to that of a Clerk of Works or Resident Engineer, and in many ways the role is similar, but it is important to recognise that those roles are usually delegated by others within their respective contracts, therefore giving the Clerk of Works or Resident Engineer little or no ultimate responsibility.

Disciplines which have carried out the Supervisor role to date include Engineers from a civil engineering, structural or process background; Architects and other design disciplines; Building Surveyors and Clerks of Works.

It is important to recognise that the Supervisor does not act on behalf of the Project Manager; it represents the Client and has its own responsibilities and obligations within the contract which may be summarised as follows:

- the Supervisor carries out and/or witnesses tests being carried out by the Contractor or some other party.
- the Supervisor may instruct the Contractor to search for a Defect.

    This is a common to other contracts where it is referred to as "opening up" or "uncovering", the principle being if no Defect is found the matter is dealt with as a compensation event.
- the Supervisor notifies the Contractor of each defect as soon as it finds it.
- the Supervisor issues the Defects Certificate.

    This is a very important action as once the Defects Certificate has been issued, the Contractor is no longer required to provide the insurances under the contract. Also dependent on which Secondary Options have been selected, any retention (Clause X16) is released back to the Contractor, if a Defect included in the Defects Certificate shows low performance, the Contractor pays low performance damages (Clause X17), and the Contractor no longer has to report performance against KPIs (Clause X20). The Defects Certificate is used on the later of the defects date and the last defect correction period, not when all defects have been corrected. In this case, the Defects Certificate may show defects which the Contractor has not corrected.

- the Supervisor marks Plant and Materials as for the contract if the contract identifies them for payment.

    Marking Plant and Materials can involve making a physical mark on them to denote that it has seen them but can include compiling an inventory, photographic records, etc. The Contractor is required to prepare the Plant and Materials for marking as required by the Scope, which can include setting them aside from other stock, protection, insurances and any vesting requirements. Once they have been marked, any title the Contractor has to them passes to the Client.

Again, as with the Project Manager, there should only be one Supervisor, who may then delegate duties and authorities to others. On a large project it is common for various personnel to be responsible for checking mechanical, electrical, structural, finishings, landscape elements, etc, but again there should be one Supervisor who delegates duties to others in respect of these elements. Each then reports back to the single Supervisor.

The Adjudicator

The Adjudicator can be chosen by one of three methods:

- By the Client specifically naming the individual in Contract Data Part 1. The Contractor therefore knows at the time of tender who the Client has named and can therefore confirm that it does not agree with the Client's choice. This has advantages in that the parties always know who will be the Adjudicator if they have a dispute. The disadvantages are firstly that parties have commented in the past that naming the individual within the Contract Data assumes that there will be a dispute, and often the Adjudicator will require payment for having its name in the contract and being available should the parties have that dispute!
- By the Client naming an Adjudicator nominating body in Contract Data Part 1. This would normally be a professional institution. This tends to be the favoured method.
- By the parties agreeing who will be the Adjudicator at the time of the dispute. The disadvantage of this method is that once the parties are in dispute they probably would not readily agree with each other on anything!

The Adjudicator becomes involved only when either contracting party refers a dispute to him. As an impartial person it is required to give a decision on the dispute within stated time limits. If either party does not accept its decision they may refer the dispute to the tribunal (litigation or arbitration). The Adjudicator's Contract requires that payment of the Adjudicator's fees is shared by the Parties unless otherwise stated.

0.4 Some unique NEC4 features and terms

Italics *and Capital Initials*

New users of the NEC contracts often become confused by the use of *italics* and Capital Initials within the contract.

This is referred to in Clause 11.1, in that terms identified in the Contract Data are in *italics* and defined terms (defined within Clause 11.2) have Capital Initials.

An example of possible confusion is Clause 11.2(3):

> "The Completion Date is the *completion date* unless later changed in accordance with the contract."

New users may question the duplication of the term "completion date", but what it means is that the Contractor completes the works by the date stated in the contract (Contract Data Part 1) . . . but this date can change either through compensation events, or acceleration.

Examples of other italicised terms include:

- *Contractor*
- *Client*
- *Project Manager*
- *works*
- *starting date*
- *defect correction period*

Examples of other Capital Initials include:

- Defect
- Early Warning Register
- Equipment
- Plant
- Scope
- Site Information

Corrupt Act

The NEC4 contracts now provide a definition of a Corrupt Act, defined in Clause 11.2(5) as:

- *the offering, promising, giving, accepting or soliciting of an advantage as an induce-ment for an action which is illegal, unethical or a breach of trust, or*
- *abusing any entrusted power for private gain in connection with this contract or any other contract with the Client.*

There is a new termination provision under Clause 91.8, the Client may terminate if the Contractor does a Corrupt Act (see also Chapter 9).

In addition, the Contractor must take action to stop a Corrupt Act of a Subcontractor or a supplier of which it is, or should be, aware. The Contractor is also required to include equivalent provisions within its subcontracts and contracts for supply of Plant and Materials and Equipment.

Equipment and Plant

The ECC definitions of Equipment and Plant are different to many other contracts.

(I)   EQUIPMENT – CLAUSE 11.2(9)

Equipment is defined in the contract as "items provided by the Contractor and used by him to Provide the Works and which the Scope does not require the Contractor to include in the works". In that sense it includes anything which is used to build the project, but not incorporated into the works, and does not remain on site after Completion.

Specific examples of Equipment would include excavators, dumpers, lorries, cranes, tools, scaffolding and temporary accommodation. These items are normally referred to as "plant" in other contracts.

It also includes associated consumables such as fuel, lubricants, tyres and other consumables.

Note that, if such items are provided by the Client for the Contractor to use then they are "equipment" not "Equipment".

(II)   PLANT AND MATERIALS – CLAUSE 11.2(14)

Plant and Materials are "items intended to be included in the works", specific examples being mechanical and electrical installations.

It is critical that NEC practitioners recognise the difference between Equipment and Plant, and not use the term "Contractor's Plant" to describe excavators, dumpers, cranes, temporary accommodation, tools, etc., as they are defined separately throughout the contract, and particularly within the Schedule of Cost Components and the Short Schedule of Cost Components when considering payments and compensation events.

Scope

Scope is defined by Clause 11.2(16) as
    "information which

- specifies and describes the Works or
- states any constraints on how the Contractor Provides the Works

and is either

- in the documents which the Contract Data states it is in or
- in an instruction given in accordance with this contract".

The Contractor Provides the Works in accordance with the Scope.
  Scope may include:

(i)   Description of the works, including drawings, specifications, etc.
(ii) Any design to be carried out by the Contractor.
(iii) Specific requirements related to Completion.
(iv) Details of other Contractors and Others who will be occupying the Working
      Areas.
(v)   Testing requirements.

Scope is considered further in Chapter 11 (Tenders)

Site Information

Site Information is defined in Clause 11.2(18) as "information which describes
the Site and its surroundings and is in the documents which the Contract Data
states it is in".
  Site Information may include:

(i)   Geotechnical information about the Site.
(ii) Information about existing buildings, structures and plant on or adjacent to
      the Site.
(iii) Environmental issues, for example, nesting birds and protected species.
(iv) References to publicly available information.
(v)   Information from utilities companies and historic records regarding plant,
      pipes, cables and other services below the surface of the site.

Site Information is considered further in Chapter 11 (Tenders)

Others

The ECC uses the term "Others" to define parties who are "not the Client, the
Project Manager, the Supervisor, the Adjudicator or a member of the Dispute
Avoidance Board, the Contractor or any employee, Subcontractor or supplier of
the Contractor".
  Essentially Others are anyone outside the contract, for example other
Contractors, statutory bodies, regulatory authorities, utilities companies, etc.

References to Others include the Contractor's obligation to cooperate with Others in obtaining and providing information, co-operating with Others and sharing the Working Areas, obtaining approval of its design from Others. Where necessary, the Contractor provides access to Plant and Materials being stored for Others notified by the Project Manager (Contractor's Obligations), the order and timing of the work of Others (Time), loss or damage to Plant and Materials supplied by Others (Risk and Insurance), if the Contractor has substantially hindered Others (Termination).

Working Areas – Clause 11.2(20)

The term "Working Areas" needs a little explanation, as some confusion has arisen amongst ECC users.

The contract defines the Working Areas under Clause 11.2(20) as those parts of the working areas which are

- necessary for Providing the Works and
- used only for work in the contract.

The Working Areas are initially the Site, the boundaries of which are defined in Contract Data Part 1, but also any additional Working Areas may be identified by the Contractor in Contract Data Part 2 and submitted as part of its tender.

An example of an addition to the Working Areas could be where the footprint of the project to be built fills the Site, in which case the Contractor may name an additional area such as an adjacent field which it proposes to use for storage or to locate its site compound. In doing so, in order to qualify as an addition to the Working Areas, the Contractor should comply with Clause 11.2(20) in that this additional area is necessary for Providing the Works and is used only for work in this contract.

The Contractor may also submit a proposal to the Project Manager for adding to the Working Areas (Clause 16.3) during the carrying out of the works. The Project Manager may refuse acceptance because the Contractor has not complied with Clause 11.2(20).

The implications of adding to the Working Areas depend on the Main Option chosen:

Options A & B	The Short Schedule of Cost Components refers to the cost of resources used within the Working Areas, so these resources if also used within the extended Working Areas would be included as cost rather than within the fee percentage when assessing compensation events.
Options C, D & E	The Schedule of Cost Components refers to the cost of resources used within the Working Areas, so these resources if also used within the extended Working Areas would be included as cost rather than within the fee percentage when assessing compensation events.

0.5 Communications

NEC4 has strict rules regarding communications under the contract, for example under Clause 13:

- each communication which the contract requires, e.g. instructions, certificates, submissions, proposals, records, acceptances, notifications, replies and other communications, must be communicated in a form which can be read, copied and recorded. This may be by electronic means, for example email, or via a project intranet, so the issue and receipt are simultaneous. Normally the parties agree a protocol for project communications.

    This requirement is particularly important in respect of instructions as the contract does not recognise verbal/oral instructions and whereas under other forms of contract a verbal/oral instruction can be confirmed by the Contractor and if not dissented from, normally within seven or fourteen days it becomes an instruction;, the NEC3 contracts do not have that provision. It is therefore incumbent upon the instructor to ensure that it is in the correct format. The instruction may be in the form of a letter, a pro forma instruction, email, or some other compliant medium.

- the Scope may require the use of a specific communication system, normally computer based. If it does, a communication has effect at the last address notified by the recipient.

- the communication has effect when it is received. If the Scope requires the use of a specific communication system as stated above, a communication has effect when it is communicated through that system.

- the Project Manager, Supervisor or Contractor replies within the period for reply stated in Contract Data Part 1, unless the contract states otherwise.

    Different periods for reply can be inserted into Contract Data Part 1 for different communications, also they can be mutually extended by agreement between the communicating parties.

- if the Project Manager is required to accept or not accept, it must state its reasons for non-acceptance. Withholding acceptance for a reason stated in the contract is not a compensation event; this is particularly in respect of:

    1   Acceptance of Contractor's design
    2   Acceptance of a proposed Subcontractor
    3   Acceptance of a Contractor's programme.

- the Project Manager issues its certificates (Payment, Completion and Termination Certificates) to the Client and the Contractor, the Supervisor issues the Defects Certificate to the Project Manager, the Client and the Contractor.

- notifications and certificates should be communicated separately from other communications, so for example, an early warning notice should not be included as part of a set of meeting minutes, or a letter which includes other subjects.

0.6 Subcontracting

The ECC has been designed on the assumption that work may be subcontracted, with Clause 26 dealing with the Contractor subcontracting work.

Note the ECC definition of a Subcontractor under Clause 11.2(19):

A Subcontractor is a person or organisation who has a contract with the Contractor to:

- construct or install part of the works
- design all or part of the works, except the design of Plant and Materials carried out by the supplier or
- provide a service in the Working Areas which is necessary to Provide the Works, except for the

  - hire of Equipment or
  - supply of people paid for by the Contractor according to the time they work

This definition has changed from NEC3 ECC, in that the previous definition included a person or organisation which may be a supplier of Plant or Materials, but has a design responsibility. This would include companies who, for example, supply roof trusses, precast concrete bridge or floor units, joinery, or even purpose made bricks because they have a design responsibility, even though they may have no responsibility for fixing the items. This has now been expressly excluded from the definition of a Subcontractor. In many contracts companies without responsibility for fixing what they supply would be classed as suppliers.

If the Contractor proposes to subcontract work it is responsible for Providing the Works as if it had not subcontracted, and also that the contract applies as if a Subcontractor's employees and equipment were its own, thus emphasising that the Contractor is wholly liable to the Client regardless of any Subcontractors he may use.

Clause 26.2 is worth specific mention as the Contractor is obliged to submit the name of each proposed Subcontractor to the Project Manager for acceptance.

Note again that the submission by the Contractor and the acceptance by the Project Manager must both be in a form which can be read, copied and recorded in order to comply with Clause 13.1.

No Subcontractor can be appointed until the Project Manager has given its acceptance, so it is vital that the Contractor submits the name of its Subcontractors as early as possible so that the Project Manager has a reasonable period in which to consider them for acceptance. The Project Manager must accept, or give its reasons for not accepting the Subcontractor within the "period for reply" stated in the contract.

A reason for the Project Manager not accepting the Subcontractor is that "the appointment will not allow the Contractor to Provide the Works". This clause probably needs a word of caution to Project Managers who may give the Contractor any one of several reasons for not accepting the Subcontractor, but if the reason is not that "its appointment will not allow the Contractor to Provide the Works" then it is a compensation event under Clause 60.1(9), "the Project Manager withholds acceptance for a reason not stated in the contract".

Examples of non-acceptance of a Subcontractor which could give rise to a compensation event are "I have never heard of your proposed Subcontractor", or "I have heard that it is not doing too well on one of your other sites", neither of which comply with the reason "its appointment will not allow the Contractor to Provide the Works".

As stated previously, some Clients have used a Z clause to change "will not allow" to "may not allow" to allow the Project Manager to express an element of doubt, without needing to justify it. The author's view is that the original wording should remain and allow the Contractor to make a judgement with the Client and the Project Manager being protected by Clause 26.1, "if the Contractor subcontracts work it is responsible for Providing the Works as if it had not subcontracted".

Whilst the contract does not state as such, clearly, if the Project Manager has concerns about a Subcontractor it should raise them with the Contractor who should be expected to provide reasonable substantiation of their competence, possibly in the form of examples of previous work and references. If, despite the Contractor providing this substantiation, the Project Manager refuses to accept the Subcontractor, then it is a compensation event.

Project Managers are often concerned about accepting a particular Subcontractor, with probably the best advice being to ask the Contractor for some substantiation and unless the substantiation proves the Project Manager's concerns, to accept the Subcontractor and remind the Contractor that under Clause 26.1 responsibility for that Subcontractor remains with the Contractor.

Whilst the contract does not require the Contractor to use an NEC contract for its Subcontractors under Clause 26.3 it is required to submit the proposed conditions of contract for its Subcontractors to the Project Manager for acceptance unless an unamended NEC contract is used, or the Project Manager has agreed that no submission is required. The reference to an unamended NEC contract is an important new provision, as in the past Contractors used the NEC Subcontract, but heavily amended in their favour!

As the Contractor cannot appoint the particular Subcontractor on the proposed subcontract conditions until the Project Manager has accepted them the Contractor should submit proposed conditions of contract as early as possible so that the Project Manager has a reasonable period in which to consider them.

A reason that the Project Manager can give for not accepting the contract conditions is that

- they will not allow the Contractor to Provide the Works or
- they do not include a statement that the parties to the subcontract act in a spirit of mutual trust and co-operation.

In addition, Options C, D, E and F also require the Contractor to submit the pricing information in the proposed subcontract documents and proposed contract data for each subcontract to the Project Manager unless the Project Manager has agreed that no submission is required.

This is an important requirement as costs arising from the subcontracts will likely become due for payment as Defined Cost.

The NEC4 Engineering and Construction Subcontract

The NEC4 Engineering and Construction Subcontract is almost identical to the ECC, providing back to back provisions, but including a number of changes appropriate to subcontracts.

GENERAL

- There is no Project Manager or Supervisor, so all actions are between the Contractor and the Subcontractor.
- Option F does not exist in the subcontract; this is because under Option F all the works are subcontracted by the Contractor, so a management contract would not be applicable between a Contractor and a Subcontractor.
- "Contract Date" and "Completion Date" are replaced by "Subcontract Date" and "Subcontract Completion Date".
- "Scope" is replaced by "Subcontract Scope", though the term "Site Information" is not changed as it still relates to "the Site".

THE SUBCONTRACTOR'S MAIN RESPONSIBILITIES

- Clause 26 related to "Subcontracting" refers to "Subsubcontracting".

PAYMENT

- The Contractor certifies a payment within two weeks of each assessment date; under the ECC, the Project Manager certified each payment within one week of each assessment date.
- Each certified payment is made within four weeks of each assessment date.

COMPENSATION EVENTS

- The compensation events are the same except that the reference to the Client, the Project Manager and the Supervisor in the ECC is replaced by the Contractor, though Clause 60.1(5), (14) and (16) include the Client, and Clause 60.1(11) refers to a test or inspection done by the Contractor or the Supervisor causing unnecessary delay.
- The Subcontractor submits quotations within one week of being instructed to do so by the Contractor, and the Contractor replies within four weeks.

RISKS AND INSURANCE

- Clause 80.1 lists the Client's and the Contractor's risks.

Where the subcontract works comprise straightforward work and impose low risks on the Contractor and the Subcontractor, the Engineering and Construction Short Subcontract may be used. Again, the NEC3 Engineering and Construction Short Subcontract is identical to the Engineering and Construction Short Contract, but includes a number of changes appropriate to subcontracts.

NOMINATED SUBCONTRACTORS

Although some other contracts provide for Nominated Subcontractors, the NEC contracts have never done so.

This is for a number of reasons:

- The Contractor should be responsible for managing all that it has contracted to do. Nomination will often split responsibilities.
- The use of Prime Cost Sums in bills of quantities means that the Contractor does not have to price that element of the work other than for profit and attendances – the more Prime Cost Sums, the less the tenderers have to price and thus there is less pricing competition between the tenderers.
- Contracts will often give the Contractor relief in the form of an extension of time in the event of a default by the Nominated Subcontractor, provided that Contractor has done all that would be reasonable or practicable to manage the default.
- Contracts will often give the Contractor relief in the form of loss and expense where it cannot be recovered from the Nominated Subcontractor, again provided the Contractor had done all that would be reasonable or practicable to manage the default. This would include the insolvency of a Nominated Subcontractor and a subsequent renomination.

0.7 Assignment

Under new Clause 28, either Party must notify the other Party if they intend to transfer the benefit of the contract or any rights under it.

In addition, the Client cannot transfer a benefit or a right if the party receiving it does not intend to act in a spirit of mutual trust and cooperation.

See the earlier comments in the Introduction about the interpretation, and difficulty of enforcing, mutual trust and cooperation provisions in a contract. In this case, how would the Client prove (or someone else disprove) that the other party intends to act in a spirit of mutual trust and cooperation? Also, one can always "intend" to do something, but later fail to comply.

0.8 Disclosure

New Clause 29 requires that the Parties must not disclose information obtained in connection with the works except when necessary to carry out their duties under the contract.

Also, the Contractor may only publicise the works with the Client's agreement.

General disclosure restrictions in contracts are very difficult for both parties to effectively control, administer and remedy. The use of a Non-Disclosure Agreement (NDA) would give the issue some focus, with specific remedies if a Party is in breach.

Appendices

Project Manager Duties

The Project Manager

GENERAL

10.1	acts as stated in the contract
10.2	acts in a spirit of mutual trust and co-operation
13.3	replies to a communication within the period for reply
to	
13.5	
13.6	issues its certificates to the Client and Contractor
13.8	may withhold acceptance of a submission by the Contractor
14.2	may delegate any of its actions
14.3	may give an instruction which changes the Scope or a Key Date
15.1	gives an early warning
15.2	prepares a first Early Warning Register
15.2	instructs the Contractor to attend a first early warning meeting
15.2	may instruct the Contractor to attend an early warning meeting
15.4	revises the Risk Register at each early warning meeting
16.2	accepts or does not accept the Contractor's proposal
17.1	notifies the Contractor as soon as aware of an ambiguity or inconsistency
17.1	notifies the Contractor as soon as aware of any illegal or impossible requirement
19.1	gives an instruction to the Contractor stating how it is to deal with the event

CONTRACTOR'S MAIN RESPONSIBILITIES

21.2   accepts or does not accept the Contractor's design

23.1   accepts or does not accept the Contractor's design of Equipment

24.1   accepts or does not accept the Contractor's replacement person

24.2   may instruct the Contractor to remove an employee

25.3   assesses the additional cost incurred by the Client for Contractor failing to meet a Key Date

26.2   accepts or does not accept the Contractor's proposed Subcontractor

26.3   accepts or does not accept the proposed subcontract documents

TIME

30.2   decides the date of Completion

31.1   receives the Contractor's first programme

31.3   accepts or does not accept the Contractor's programme

32.2   accepts or does not accept the Contractor's revised programme

34.1   may instruct the Contractor to stop or not to start any work

35.3   certifies the date upon which the Client takes over a part of the works

36.1   instructs the Contractor to submit a quotation for acceleration

TESTING AND DEFECTS

40.2   accepts or does not accept the Contractor's quality policy statement and quality plan

41.6   assesses the cost of repeating a test after a Defect is found

44.4   arranges for the Client to allow the Contractor access to any part of the works

45.1   may change the Scope to accept a Defect

45.2   accepts or does not accept the Contractor's quotation for accepting a Defect

46.1   assesses the cost of having a Defect corrected by other people

46.2   assesses the cost to the Contractor of correcting the Defect

PAYMENT

50.1	assesses the amount due at each assessment date
50.2	considers any application for payment from the Contractor
50.6	corrects any wrongly assessed amount due
51.1	certifies a payment within one week of each assessment date
53.1	makes an assessment of the final amount due

COMPENSATION EVENTS

60.1(17)	notifies a correction to an assumption
61.1	notifies the Contractor of a compensation event
61.3	is notified by the Contractor of a compensation event
61.4	notifies the Contractor of its decision that the Prices, the Completion Date and the Key Dates are not to be changed
61.5	notifies the Contractor that it did not give an early warning
61.6	states assumptions about an event too uncertain to be forecast reasonably
62.1	may instruct the Contractor to submit alternative quotations
62.3	replies within two weeks of receipt of a Contractor's quotation
62.4	instructs the Contractor to submit a revised quotation
62.5	extends the time allowed for submitting or replying to a quotation
62.6	replies to the Contractor's notification
63.2	may agree with the Contractor rates and lump sums to assess the change in Prices
63.7	assesses a compensation event as if the Contractor had given early warning
63.11	corrects the description of the Condition for a Key Date
64.1	assesses a compensation event
64.2	assesses the programme for remaining work
64.3	notifies the Contractor of its assessment of a compensation event
64.4	replies to the Contractor's notification
65.1	may instruct the Contractor to submit a quotation for a proposed instruction
65.1	implements each compensation event by notifying the Contractor

52.3	is allowed to inspect the Contractor's accounts and records
54.1	assesses the Contractor's share
54.3	makes a preliminary assessment of the Contractor's share at Completion
54.4	makes a final assessment of the Contractor's share
93.4	assesses the Contractor's share after certifying termination

Option D

11.2(26)	decides Disallowed Cost
20.3	receives advice from the Contractor on practical implications of design and subcontracting arrangements
20.4	prepares with the Contractor forecasts of total Defined Cost for the works
26.4	accepts or does not accept the Contractor's proposed pricing information
52.3	is allowed to inspect the Contractor's accounts and records
54.5	assesses the Contractor's share
54.7	makes a preliminary assessment of the Contractor's share at Completion
54.8	makes a final assessment of the Contractor's share
60.6	gives an instruction to correct a mistake in the Bill of Quantities
93.5	assesses the Contractor's share after certifying termination

Option E

11.2(26)	decides Disallowed Cost
20.3	receives advice from the Contractor on practical implications of design and subcontracting arrangements
20.4	prepares with the Contractor forecasts of total Defined Cost for the works
26.4	accepts or does not accept the Contractor's proposed pricing information
52.4	is allowed to inspect the Contractor's accounts and records

Option F

11.2(27)	decides Disallowed Cost
20.3	receives advice from the Contractor on practical implications of design and subcontracting arrangements
20.4	prepares with the Contractor forecasts of total Defined Cost for the works
26.4	accepts or does not accept the Contractor's proposed pricing information
52.4	is allowed to inspect the Contractor's accounts and records

SECONDARY OPTIONS

X4	accepts or does not accept a proposed alternative guarantor
X10	gives an early warning
	accepts or does not accept a first Information Execution Plan
X13	accepts or does not accept a performance bond
X14	accepts or does not accept a bank or insurer
X16	accepts or does not accept a bank or insurer
X20	receives Contractor's report against KPIs
X20	receives Contractor's proposals for improving performance
X21	receives Contractor's proposal that the Scope is changed and consults with the Contractor
X22	accepts or does not accept the Contractor's forecasts of total Defined Cost issues or does not issue a notice to proceed to Stage Two
	discusses with the Contractor ways of dealing with the change to the Budget
	makes a preliminary and final assessment of the budget incentive
Y(UK)1	accepts or does not accept details of the banking arrangements receives the Authorisation

Supervisor duties

The Supervisor

10.1	acts as stated in the contract
10.2	acts in a spirit of mutual trust and co-operation

13.3	replies to a communication within the period for reply
to	
13.5	
13.6	issues certificates to the Project Manager, the Client and the Contractor
14.2	may delegate any of its actions
41.3	notifies the Contractor of tests and inspections
41.5	does its tests without causing unnecessary delay
42.1	notifies the Contractor that Plant and Materials have passed tests
43.1	may instruct the Contractor to search for a Defect
43.2	notifies the Contractor of each Defect which it finds
44.3	issues the Defects Certificate
71.1	marks Equipment, plant and Materials outside the Working Areas

1 Early warnings

1.1 Introduction

Early warnings are covered within the ECC by Clauses 15.1 to 15.4. They are also mentioned in Clauses 61.5 and 63.7, which we will discuss later.

Early warnings are a key component of the overall risk management process in the ECC. The process is not about liability, but instead about the Parties collaborating to identify, mitigate or remove the effect of matters which could cause difficulty.

The ECC obliges the Project Manager and the Contractor to notify each other as soon as either becomes aware of any matter which could affect the project in terms of time, cost or quality.

The requirement to notify must be done in a form which can be read, copied and recorded, and separately from other communications, in accordance with Clauses 13.1 and 13.7. The obligation is to notify as soon as either becomes aware of a matter, and this can often be difficult for parties to demonstrate one way or the other.

Whilst there is no mandatory format for an early warning, Figure 1.1 shows an example which complies with the contractual requirements.

Sometimes correspondence or records may show when the Contractor or Project Manager first became aware of something, but this can be a matter of subjective interpretation. Note that the contract says "*becomes aware*" not "*should have become aware*".

This sometimes causes confusion, but if we take, as an example, where the Contractor is instructed by the Project Manager to use a particular methodology and the Contractor knows from experience that the methodology would probably not be sufficient to meet the requirements of the Scope, then the Contractor should give an early warning.

Under Clause 15.1 the Contractor and the Project Manager give an early warning by notifying the other as soon as either becomes aware of any matter which could:

Contract:	EARLY WARNING NOTICE
Contract No:	EWN No...

Section A: Enquiry

To: Project Manager/Contractor

Description

This matter could:

- ☐ Increase the total of the Prices
- ☐ Delay Completion
- ☐ Delay meeting a Key Date
- ☐ Impair the performance of the works in use

| Early warning meeting arranged | Yes/No | Date: .... |

Signed:...............................(Contractor/Project Manager)   Date: ....

Action by: Date required:

Section B: Reply

To: Contractor/Project Manager

Signed:...............................(Contractor/Project Manager) | Date: ....

Copied to:

Contractor ☐   Project Manager ☐   Supervisor ☐   File ☐   Other .... ☐

Figure 1.1 Example of Early Warning Notice

- increase the total of the Prices

  The price of the works, in the form of the activity schedule, the bill of quantities, or the target.
- delay Completion

  The completion of the whole of the works.
- delay meeting a Key Date

  The completion of an intermediate "milestone date" in accordance with Clause 11.2(11).
- impair the performance of the works in use.

  This sometimes causes confusion, but if we take, as an example, where the Contractor is instructed by the Project Manager to use a particular type of pump and the Contractor knows from experience that that pump would probably not be sufficient to meet the Client's requirements once the works are taken over, then the Contractor should give early warning.

The Contractor may also give an early warning by notifying the Project Manager of any other matter which could increase its total cost. One could query whether a matter which could increase the Contractor's cost, but not affect the Prices, should be an early warning matter, or for that matter, whether it should be anything to do with the Project Manager, particularly if Option A or B has been selected.

However, the words are *"the Contractor may give an early warning"* so it is not obliged to do so. This provision is designed to encourage collaboration between the parties, irrespective of their contractual liability.

Note, also within Clause 15.1, the Contractor is not required to give an early warning for which a compensation event has previously been notified, so, as an example, if the Project Manager gives an instruction which changes the Scope, it is a compensation event, for which neither the Project Manager or the Contractor are required to give early warning.

The early warning procedure obliges people to be "proactive", notifying and dealing with risks as soon as the parties become aware of them, rather than "reactive", waiting to see what effect they have, then trying to deal with them when it is often too late. Encouraging the early identification of problems by both parties, puts the emphasis on joint solution finding rather than blame assignment and contractual entitlement.

It is one of the most important and valuable aspects of the NEC4 contracts and it is perhaps surprising that, whilst few other contracts refer to an early warning process, only the NEC4 contracts set out in clear detail what the parties are obliged to do, with appropriate sanctions should the parties not comply (see Clauses 11.2(8), 61.5 and 63.7).

The author has encountered situations where Clients and Project Managers appear hostile to the receipt of early warnings from Contractors. Sometimes they are viewed as the first stage in a compensation event process or similar. That may be so, but not always, and in effect could prevent a compensation event occurring or at least lessen its effect.

Statistically speaking, one would expect an equal number of early warnings to be issued by the Project Manager and the Contractor. Typically though, those issued by the Contractor frequently outnumber those issued by the Project Manager.

There are sanctions for the failure to issue early warnings and we will discuss these later. There is no "ready reckoner" for how many early warnings should be issued but an absence of them should be viewed as more worrying than a plethora of them.

1.2 Notifying early warnings

The contract requires (Clause 13.1) that all communications, for example instructions, notifications, submissions, etc. are in a form that can be read, copied and recorded, so early warnings should not be a verbal communication, such as a telephone conversation. If the first notification is a telephone conversation, or a comment in a site meeting, it should be immediately confirmed in writing to give it contractual significance.

Also, Clause 13.7 requires that notifications which the contract requires must be communicated separately from other communications, therefore early warnings must not be included within a long letter which covers a number of issues, or embodied within the minutes of a progress meeting.

There are some key words within the obligation to notify:

- *"The Contractor and the Project Manager"* – no-one else has the authority or obligation to give an early warning. The Project Manager is therefore notifying on behalf of itself, the Client, the Supervisor, the Client's Designers and many possible others who it represents within the contract. The Contractor is notifying on behalf of itself, its Subcontractors, its Designers (if appropriate), and again many possible others who it represents under the contract. Early warnings should be notified by the key people named in Contract Data Part 2.

    Project Managers are often criticised for seeing early warnings as something the Contractor has to do, and in fact most early warnings are actually issued by the Contractor. However, the Contractor and the Project Manager are obliged to give early warnings each to the other, so it is critical that Project Managers play their part in the process.

    As an example, if the Project Manager becomes aware that it will be late in delivering some design information to the Contractor, it should issue the early warning as soon as it becomes aware that the information will not be delivered to the Contractor, not wait and subsequently blame the Contractor for not giving an early warning stating that it has not received the information!

- "As soon as" – means immediately. There are a number of clauses within the contract that deal with the situation where the Contractor did not give an early warning. Whilst the party who gives the early warning must do so as soon as it becomes aware of the potential risk, the other party should respond as soon as possible and in all cases within the period for reply in Contract Data Part 1.

- "Could" – not must, will or shall. Clearly there is an obligation to notify even if it is only felt something may affect the contract, but there is no clear evidence that it will.

The Project Manager enters early warning matters in the Early Warning Register. Early warning of a matter for which a compensation event has previously been notified is not required.

It must be emphasised that early warnings are not the first step toward a compensation event as is often believed. Early warnings feature in a completely separate section of the contract and in fact the early warning provision is intended to prevent a compensation event occurring or at least to lessen its effect. It can also be used to notify a problem which is totally the risk of the notifier. It is also worth mentioning that early warnings are a notice of a future risk, not a past one. The parties are not required, nor is it of any value, to notify a risk that has already happened.

The nature and format of early warnings can have an impact on how they are received and what reaction is prompted. Contracting parties are often keen to identify the flaws of their counterparties, but the obligation here potentially requires parties to identify and record their own failures. Common sense is therefore needed in how these are identified.

The "matters" which require an early warning to be given are essentially the three determinants of success in any project, price, time and quality. In management terms these three are inextricably linked and will always be measured by clients. Therefore, the contract recognises the importance of managing "matters" that may impact on these.

Example

Two weeks before the completion date the Contractor is informed by its Subcontractor that they have insufficient ceiling tiles to complete the works and that this could potentially cause a delay to Completion.

Should the Contractor issue an early warning to the Project Manager, bearing in mind that there has been a failure on the part of the Subcontractor, and probably the Contractor?

Yes, it should issue an early warning as it could delay Completion. Of course, there is no entitlement on the part of the Subcontractor or the Contractor in terms of a compensation event, but it should still issue the early warning.

The Project Manager and Contractor may possibly hold an early warning meeting, also inviting the Subcontractor to attend and they could consider proposals and seek solutions, and decide action such as the Client taking over the works and the Contractor and Subcontractor returning to site when the material has been delivered, or the Client taking over the parts of the works which are not related to this particular material. With a degree of innovation and collaboration brought about by the early warning provision the effect of this oversight could be mitigated.

The first bullet in 15.1 refers to a defined term, "Prices". The definition of this term depends upon which Main Option has been selected, as it is a function of the payment risk allocated between Client and Contractor. The definitions are in Clauses 11.2(32), 11.2(33) and 11.2(34). Where the matter increases the Prices, an early warning must be given. Note that there is not a corresponding obligation to notify something that reduces the total of the Prices.

The second and third bullets clearly require reference to the latest Accepted Programme.

The fourth bullet in Clause 15.1 is more subjective and relates to the quality of the completed work. One of the skills necessary in operating these provisions is the ability to judge how material a risk should be before it becomes the subject of an early warning. Minor risks are probably discussed, and resolved, by representatives of the Contractor and the Project Manager every day of the week but do not need to become the subject of early warnings. Many conversations are effectively short early warning meetings, but would never be formally categorised as such. Such minor matters would not be worth pursuing through Clause 15. However where the use of an early warning would remove or mitigate the risk of an impact upon the project delivery, the early warning process should be used.

Clause 15.1 also contains an optional action for the Contractor, a rare item in NEC contracts. The Contractor *may* give an early warning of something that could increase its total cost. This is different from the absolute obligation to give an early warning of something that could increase the total of the Prices. It sits comfortably also with the requirements in Clause 20.4 for the Contractor to provide cost forecasts to the Project Manager. Where the Contractor chooses to give an early warning in this scenario then it is likely that it is seeking assistance for a matter that at least affects him, but possibly both Parties.

Early warnings are a proactive, not reactive, mechanism and, as stated previously, it is pointless issuing an early warning of a matter that has passed.

A schedule of early warnings (see Figure 1.2) should be kept, identifying and numbering each early warning, with additional information about each one.

The author has found the schedule useful for tracking early warnings as the project progresses, but also at the end of each project to review the schedule and each early warning notified during the life of the project as "lessons learned" to consider in drafting Scopes for future contracts.

1.3 Remedy for failure to give early warning

One of the objectives of all NEC contracts is that the contract's provisions should act as a stimulus to good management. Typically that is as a financial incentive for doing what the contract requires or, more often, a sanction for the failure to comply. Early warnings are no exception. There are sanctions within the ECC for the failure of either the Project Manager or the Contractor to comply with these provisions.

There are express provisions covering the Contractor's failure to issue an early warning when it should have done, namely in Clauses 61.5 and 63.7. Clause 63.7 often confuses people on first reading, appearing to provide no sanction for the failure. The reverse is true. The Contractor, having failed to give

Contract: ....
Contract No: ....

SCHEDULE OF EARLY WARNINGS

Ref.	Date notified	Origin? Project Manager or Contractor	Description of event	Relevant Clauses	Action with PM or Contractor	Date Resolved	Notes
1							
2							
3							
4							
5							
6							
7							
8							
9							
10							

Figure 1.2 Example of Early Warning schedule

an early warning, will only gain the compensation that would have accrued from the (presumably) lower cost resolution of the matter that would have followed the giving of an early warning.

Users of the NEC contracts often query the lack of an express sanction for the failure of the Project Manager to give an early warning when it should have done so. Without looking at the contract's provisions it is reasonably apparent that such a failure will, in many circumstances, lead to a cost increase or other loss for the Client and therefore the Project Manager should be incentivised to comply.

Considering Clauses 10.1, 10.2 and 15.1 together we can see that the Project Manager is obliged to give early warnings of matters that it is aware of. Failure of the Project Manager to do this could lead to a breach of contract by the Client and this will be a compensation event under Clause 60.1(18).

Early warnings often lead to an early warning meeting and Clause 15.2 allows the Project Manager and the Contractor to instruct the other's attendance at the meeting. Some commentators argue that, in certain legal jurisdictions, there is an implied obligation for the Client to attend if instructed. Failure of either Party to attend would be a breach of the contract and would lead to the sanctions described above.

As stated above, under Clauses 61.5 and 63.7, if the Project Manager decides that the Contractor did not give an early warning of the event which an experienced contractor could have given, it notifies this to the Contractor when it instructs him to submit quotations. If the Project Manager has done so, the event is assessed as if the Contractor had given early warning.

Example

On 1 February, the Contractor becomes aware of an issue which could increase the total of the Prices and delay Completion. It should immediately give an early warning to the Project Manager; however, it fails to do so until 22 February, three weeks later.

When the Contractor gives the early warning, the Project Manager notifies a compensation event and instructs the Contractor to submit a quotation, at the same time notifying the Contractor that it did not give an early warning of the matter that an experienced Contractor could have given. If the Project Manager has given such notification, then the compensation event is assessed as if the Contractor had given early warning, so the Contractor should price its quotation based on the event as at 1 February, rather than 22 February.

Failure to give an early warning can also be considered as Disallowed Cost under Options C, D, E and F, Disallowed Cost being cost which the Project Manager decides was incurred only because the Contractor did not give an early warning which the contract required him to give.

1.4 Dealing with risk and liabilities

All commercial contracts are concerned with risk and with the allocation of risks to the Parties. NEC contracts go one step further than many other contracts and provide procedures for directly managing risk. In line with the NEC's objectives, the Parties are incentivised to follow those procedures. The early warning process is concerned with applying foresight to likely matters that could affect the delivery of the project to the end users and the Client.

Clause 80.1 defines the Client's liabilities, and these are relatively standard risks taken by Clients in construction contracts (see Chapter 8).

There is an entry in Contract Data Part 1 where the Client may list additional Client's liabilities. This entry of the Contract Data *does* amend the risk allocation of the contract, unlike the two entries which we describe later when discussing the Early Warning Register. There is not a corresponding section in Contract Data Part 2 for the Contractor to do likewise, although this could be facilitated by appropriate provisions in a tender procedure if the Client wanted to solicit such ideas.

Dependent on the Main Option chosen, a Contractor's liability may still be paid for by the Client, even though it will not be a compensation event.

Clause 81.1 then states the Contractor's liabilities.

By Clause 82, the Parties agree to indemnify one another against those liabilities. It is no coincidence that the immediately following clause relates to insurance. The Contractor's ability to indemnify the Client against the risks will most probably be underwritten by insurance.

Where a Client's liability occurs and this becomes a compensation event (Clause 60.1(14)), its impact is assessed in accordance with the compensation event provisions.

Liabilities can be classified in an ECC contract as:

(i) Those which are owned by the Client and do not affect the Contractor.
(ii) Those which are owned by the Client and do affect the Contractor in terms of Price, Key Dates and/or Completion Date, and are therefore included and managed as compensation events.
(iii) Those which are owned by the Contractor and are therefore deemed to have been included within its price and programme. These are risks which may affect the Contractor but are not listed as compensation events.

1.5 General principles of risk management

There are a number of definitions of risk, an example being:

> "*Any uncertain event or set of circumstances that, should it or they occur, would have an effect on achieving the objectives.*"

> (Association for Project Management)

The objectives in a construction project are to build safely and to the required quality, to complete within the required time scale, and within the required budget.

Risk is always inherent in any activity, no matter how simple or complex it may be. However, the degree of risk will vary depending upon the circumstances involved and the amount of risk that one is willing to assume is normally directly proportional to the amount of reward that is anticipated. The greater the anticipated reward, the greater the risk one is normally willing to assume.

Risk is an integral part of the construction process, and an important factor in the determination of construction costs and durations.

The parties to a construction contract assume that the contract will not only provide them with adequate protection, but also provide a fair and workable framework within which a vastly complex project can be constructed to the mutual satisfaction of all concerned.

The construction contract itself should be clear, concise, complete, legally correct, and equitable. It should address the risks directly and attempt to eliminate as much of the ambiguity as possible.

Characteristics causing construction projects to present risks

All construction projects can be regarded as having risk as an underlying factor, though there are a number of differences between a construction contract and a manufacturing process, in that, every construction project:

(i) is different, the only similarities being the people who manage them and the materials and equipment used.
(ii) is not produced in a factory environment, but on a site which is subject to fairly predictable seasonal variations, but unpredictable weather exceptions.
(iii) brings a temporary multi skilled and experienced team together which is formed for the purpose of constructing the project and whose members may possibly never work together again.

A thought provoking quotation, while written some time ago and in a "comical" style, still holds a certain degree of truth:

> ". . . the so called project team. As teams go it really is rather peculiar, not at all like a cricket eleven, more like a scratch bunch consisting of one batsman, one goal keeper, a pole vaulter and a polo player. Normally brought together for a single enterprise, each member has different objectives, training and techniques and different rules. The relationship is unstable, even unreliable, with very little functional cohesion and no loyalty to a common end beyond that of each member coming through unscathed."

> Denys Hinton, 1976

As a result of the above, one can easily draw the conclusion that:

(i) It is impossible to produce a design which will not require subsequent modifications, either because of client changes or because the design is found in practice not to work.

(ii) It is impossible for a contractor to produce a tender which will correspond exactly to the cost plus overheads and profit which will actually be incurred during the construction.

(iii) It is impossible for a contractor to produce a programme which can in all respects apply without revision and updating.

(iv) When people are faced with such uncertainty they may, in an effort to defend their own beliefs and organisations, engage in conflict situations.

As a consequence it is clear to see that both client and contractor organisations are faced with risks that occur as a result of the nature of construction.

From the moment the decision to begin to redesign is taken until the new facility is in use, the client and contractor take on board risks.

In addition to inherent risks within the construction risks there are also risks that will arise from the agreement between the parties. These contractual risks are primarily expressed within the contract as intended between the parties, but they can also arise through inexperienced drafting which can lead to lack of contract clarity, ambiguities and inconsistencies.

The main principles of apportioning risk in construction contracts areas follows:

- The client will always pay for risk, either by retaining the risk within the contract and paying for it should it occur, or the Contractor prices it within its tender.
- They should be apportioned to the party best able to manage and/or control them and to sustain the consequences if risks materialise.
- Risks outside the contractor's control and/or foreseeability should remain with the client.
- The party carrying the risk should be motivated to manage the risk, possibly through a risk/reward process.

The major types of risks that can arise on construction projects follow.

Risks to Client

Technical

- Ability of existing design to deliver sufficient standard and quality
- Incomplete or incorrect design
- Inadequate site investigations

Logistical

- Availability of resources
- Ability to avoid business interruption
- Availability of transportation

Resources

- Insufficient resources and expertise
- Weather and seasonal problems
- Industrial relations problems

Timing

- Inability to give access or possession to the Contractor

Financial/contractual

- Inflation
- Delay in payment
- Non-availability of funds
- Cash flow problems
- Ambiguities and/or inconsistencies in contract documents
- Work carried out by directly employed parties
- Losses due to default of Contractors

Safety/reliability

- Delays due to having to satisfy requirements

Quality assurance/standards

- Quality issues

Choice of site/physical

- Ground conditions
- Ground water conditions
- Exceptional weather conditions
- Damage to neighbouring buildings
- Soil strata type and variability
- Ground water conditions
- Contamination

Environmental and planning

- Existing rights of way
- Public enquiries
- Planning requirements
- Noise abatement
- Availability of services

Political

- Changes in government/new legislation
- Customs and import restrictions

Design

- Suitability and adaptability of the design

Unfamiliarity

- Pioneer design
- Experience of design team

Inflation

- Differential inflation due to market situation at time of tendering
- Abnormal inflation

Errors in tender documents

Construction

- Bankruptcy/Insolvency
- Industrial Action

Variations

- Effect on construction duration and price of late variations
- Effect of number of orders

Construction delays

- Delay by specialist Contractors and statutory authorities
- Delay in giving instructions to the Contractor

Force majeure

- Fundamental risk, which cannot be priced

The risk management process

The allocation of clients' risks and contractors' risks reflects the parties' allocation at the commencement of the contract. As events arise during the delivery of the project, the contract will allocate responsibility and will sometimes define a process for the parties to follow.

In the ECC, early warnings are one of those processes, focused on the resolution of the problem rather than the blame or allocation of liability. The author has seen repeated successful projects from parties who understand the role and purposes of the early warnings, Early Warning Register and early warning meetings. These three tools dovetail with one another and act as a benefit to parties who embrace them.

Risk management is a structured approach to identifying, assessing and managing risk.

It can be defined as:

"The systematic application of management policies, procedures and practices to the tasks of identifying, analysing, assessing, treating and monitoring risks".

The risk management process can help a party analyse the degree of uncertainty in achieving the contract/project objectives, understand the consequences of particular events occurring, and develop management actions to control the event or minimise its consequences.

There are six primary activities in the process:

1 Risk Identification

Risks should be identified as early as possible and continuously monitored to highlight new risks and the passing of old risks. Risk identification should initially be carried out by analysing all available documentation and knowledge as a brainstorming session involving a number of individuals, ideally chaired by an independent person external to the project.

2 Risk Analysis

Once the risks have been identified they should be classified in terms of:

- likelihood of occurring
- impact on the project objectives should they occur.

Each risk can then be given a ranking, possibly:

- Extremely unlikely
- Unlikely
- Moderate
- Likely
- Almost certain

Impact can also be given a ranking, possibly:

- Insignificant
- Moderate
- Major
- Severe
- Catastrophic

3 Risk Planning

This is the process of developing responses to address, manage or control risks.

The responses will consider:

- **Avoiding the risk** – the client abandoning the project, the contractor deciding not to tender.
- **Reducing the probability** – doing something a different way or educating the people that may be subject to the risk, so the risk is less likely to occur.
- **Reducing the impact** – including resources, funds and/or float in programmes to cushion the effect of the risk.
- **Transferring the risk** – passing the risk to another (willing) party or an insurance company who is best able to manage or bear the risk.
- **Accepting the risk** – with the full knowledge of what the risks and potential outcomes are.
- **Ignoring the risk** – with the full knowledge of what the risks and potential outcomes are.

4   Risk Tracking

Risks should be tracked throughout the project to monitor the status of each risk as the project progresses.

5   Risk Controlling

An activity that utilises the status and tracking information about a risk or risk mitigation effort, a risk may be closed or watched, a contingency plan may be involved.

6   Risk Communicating

An action to communicate and document the risk at all times. This will normally be done in the form of a risk register.

When considering risk it is vital to consider:

- Which party can best foresee the risk
- Which party can best bear the risk
- Which party can best control the risk
- Which party most benefits or suffers if the risk materialises.

1.6 The use of an Early Warning Register under the ECC

The NEC4 contracts provide for the use of a risk register, the revised contracts now referring to it as an "Early Warning Register".

The Contract Data does contain matching sections in Parts 1 and 2 for the Parties to add matters that will be included in the Early Warning Register. That is to say, both Parties have the ability to list risks that they wish to form part of the risk management processes in the contract. These, as we will see, **do not** change the risk allocation between the Parties. These entries are to assist the risk management processes of the Parties by listing those risks which collaborative behaviour will help with.

It is critical that the parties fully understand that the purpose of an Early Warning Register is to list all the identified risks and the results of their analysis and evaluation. It can then be used to track, review and monitor risks as they arise to enable the successful completion of the project. The Early Warning Register does not allocate risk that is done by the contract.

The Early Warning Register is a simple document, but a vital one in the process of risk management. Its role and contents are frequently misunderstood by Parties, particularly where people confuse its use with risk registers elsewhere, such as those that are components of company management systems. Its purpose is to assist the Parties and the Project Manager with managing risk in the project delivery and not to allocate responsibility or blame.

It is helpful to start with its definition from Clause 11.2(8);

The Early Warning Register is a register of matters which are

- *listed in the Contract Data for inclusion and*
- *notified by the Project Manager or the Contractor as early warning matters.*

It includes a description of the matter and the way in which the effects of the matter are to be avoided or reduced.

So, reading the definition carefully, the Early Warning Register contains information about these risks;

- Those listed in the Contract Data

  - by the Client in Contract Data Part 1
  - by the Contractor in Contract Data Part 2

- Those notified as an early warning matter

  - by the Project Manager
  - by the Contractor

The information contained in the Early Warning Register is:

- a description of the risk
- the actions to be taken to avoid or reduce the risk.

The Early Warning Register need not contain anything else. It is a document produced after the Contract Date and therefore it does not form part of the contract.

For practical reasons though it should not contain any statements that purport to allocate risk to one party or the other, that will just prove confusing in the long run.

The contract does not prescribe the format or layout of the Early Warning Register, other than to list the risks in the contract and those that come to light at a later date and are notified as early warnings following which, if there is an early warning meeting, the Project Manager revises the Early Warning Register to record the outcome of the meeting.

Note that by the definition within Clause 11.2(8), risks which were not originally included in the Contract Data or subsequently notified as early warnings should not be included in the Early Warning Register. Contract Data Part 2 allows the Contractor to identify matters which will be included in the Early Warning Register. The Early Warning Register does not allocate or change the risks in the contract, it records them and assists the parties in managing them. In that sense it is a valuable addition to the workings of the contract.

In effect, the Contractor could seek to qualify its tender by including additional matters within Contract Data Part 2 which it sees, or would like to transfer as Client's liabilities. That is not the intent of the ECC as the only place to amend the risk profile is in Contract Data Part 1. However, the wording of the final bullet of Clause 80.1 leaves open the possibility of the Contractor inserting such an entry into Contract Data Part 2. Clients should check for such entries during tender appraisal.

In order to compile the Early Warning Register the risks must first be listed, then they are quantified in terms of their likelihood of occurrence and their potential impact upon the project.

In that sense, the contract does not prescribe the format or layout of the Early Warning Register, or its intended purpose, other than to list the risks in the contract and those that come to light at a later date and are notified as early warnings following which, if there is an early warning meeting the Project Manager revises the Early Warning Register to record the outcome of the meeting.

Note that by the definition with Clause 11.2(8), risks which were not originally included in the Contract Data or subsequently notified as early warnings should not be included in the Early Warning Register. Contract Data Part 2 allows the Contractor to identify matters which will be included in the Early Warning Register.

The Early Warning Register does not allocate or change the risks in the contract, it records them and assists the parties in managing them. In that sense it is a valuable addition to the contract which was absent from previous editions but introduced as a Risk Register in NEC3.

Figure 1.3 shows an extract from a typical register which would comply with the contract in that it includes the basic requirement for a description of the risk and a description of the actions which are to be taken to avoid or reduce the risk.

There is no stated list of components of an Early Warning Register, but column headings should typically be titled:

(i) Description of risk

A clear description of the nature of the risk, if necessary referring to other documents such as site investigations, etc.

(ii) Implications

What would happen if the risk were to occur?

(iii) Likelihood of occurrence

This provides an assessment of how likely the risk is to occur. The example shows a forecast on a 1 (least likely) to 5 (most likely) basis, though it may be assessed as percentages, colour coding, or simply "Low" (less than 30% likelihood), "Medium" (31–70% likelihood), "High" (more than 70% likelihood).

EARLY WARNING REGISTER

Contract:............

Contract No:............

Contractor:............

Project Manager............

Description of Risk	Implications	Likelihood of Occurrence (1–5) (Least – Most)	Potential Impact (1–5) (Low – High)	Risk Score	Risk Owner	Mitigation Strategy By whom? By when?	Allowance in the total of the Prices	Programme Allowance	Client cost allowance	Risk Status	Last Updated
Discovery of unforeseen existing silo bases	Delay and additional costs in removal	2	4	8	Initially Contractor unless additional silo bases discovered then Client as comp. event	Site Information identifies likely locations Contractor to take due regard	No allowance for unforeseen	No allowance for unforeseen	£20,000	Reducing Excavation 50% complete No additional silo bases discovered to date	15/02/2018

Figure 1.3 Example of Early Warning Register

(iv) Potential impact

This assesses the impact that the occurrence of this risk would have on the project in terms of time and/or cost. The example shows the assessment on a 1 (low) to 5 (high) basis.

(v) Risk score

Risk = Likelihood of something occurring x Impact should it occur.

By evaluating risk on that basis each can be evaluated and categorised into an order of importance. This formula can only be applied to economic loss, and different standards would need to be adopted when considering risk of death or serious bodily injury.

(vi) Risk owner

This identifies which party or individual owns the risk; this is allocated through the contract, the principle being that if the Contractor owns it, it is deemed to have allowed for its price and/or time effect within its tender. If the Client owns it, it will be a compensation event.

(vii) Mitigation strategy

This registers what actions are proposed that could be taken to prevent, reduce, or transfer the risk. Also, who is responsible for this action, and when should the action take place?

(viii) Allowance in the total of the Prices

How much has the Contractor allowed in its tender for a risk it owns?

(ix) Programme allowance

How much time has the Contractor allowed in its programme for a risk it owns?

(x) Client cost allowance

If it is a Client's risk, how much has it allocated to cover a potential compensation event?

(xi) Risk status

This identifies whether the risk is current, and also whether it is increasing, decreasing, or has not changed since it was last reviewed?

(xii) Last updated

When was an additional risk identified? When was the Early Warning Register last updated?

In effect, the Contractor could seek to qualify its tender by including additional matters within Contract Data Part 2 which it sees, or would like to transfer, as Client's risks.

Risk management is essential to the success of any project and normally follows five steps:

Step 1 – Identify the risks

Step 2 – Decide who could be harmed and how

Step 3 – Evaluate the risks and decide on precautions

Step 4 – Record the findings and implement them

Step 5 – Review the assessment and update if necessary

In order to compile the Early Warning Register the risks must first be listed, then they are quantified in terms of their likelihood of occurrence and their potential impact upon the project.

1.7 Early warning meetings within the ECC

The Early Warning Register becomes more important once early warnings lead to early warning meetings, see Clauses 15.2, 15.3 and 15.4.

The Project Manager prepares a first Early Warning Register and issues it to the Contractor within one week of the starting date.

The Project Manager also instructs the Contractor to attend a first early warning meeting within two weeks of the starting date and further early warning meetings at no longer than the durations stated in the Contract Data, until Completion of the whole of the works.

Clause 15.2 includes a rare example of an optional action in an NEC contract, when, after the first early warning meeting, either the Project Manager or the Contractor may instruct the other to attend an early warning meeting.

Note the key word in Clause 15.2 is "instruct". If the Contractor calls an early warning meeting, it is instructing, not merely requesting, the Project Manager to attend. This clause is clearly intended to promote ownership of the project by the Contractor and the Project Manager, and any issues that could affect it, together with their resolution. Hopefully in the real world such meetings are arranged by co-operation and invitation.

Notwithstanding that the Project Manager or the Contractor can instruct the other to attend an early warning meeting, early warning meetings are also held on a periodic basis no longer than the interval stated in the Contract Data until Completion of the whole of the works.

An early warning meeting may only have the Project Manager and the Contractor present, though each may instruct other people to attend if the other agrees; also a Subcontractor may attend, so in reality a number of people normally attend the meeting.

On a lighter note, the "rule of meetings" will often apply in that the productivity of the meeting is often inversely proportional to the number of people who attend it! However, this should not mask the importance of getting the right people to attend. Project delivery, particularly in buildings, relies on multiple specialists and these should be encouraged to participate wherever necessary.

Clause 15.3 describes what the attendees at an early warning meeting should do in an effort to either resolve the problem or at least attempt to resolve it.

- making and considering proposals for how the effect of each matter in the Early Warning Register can be avoided or reduced

- seeking solutions that will bring advantage to all those who will be affected
- deciding on the actions which will be taken and who, in accordance with this contract will take them
- deciding which matters can be removed from the Early Warning Register and
- reviewing actions recorded in the Early Warning Register and deciding if different actions need to be taken and who, in accordance with the contract, will take them.

Clearly the purpose of the early warning meeting is to actively consider ways to avoid or reduce the effect of the matter which has been notified. In some cases, the matter can be fully resolved, but in others, as the matter may not yet have occurred it may simply have to be "parked" and recorded as such in the Early Warning Register. It is the Project Manager's responsibility to revise the Early Warning Register to record the outcome of the meeting, and to issue the revised Early Warning Register to the Contractor.

Early warning meetings can happen in several ways:

- As a standing agenda item in regular (say weekly, or monthly) meetings
- As a routine meeting (say once a month)
- Ad hoc, when the matter is sufficiently urgent.

Meetings can provide a distraction to the "day job", so complying with Clause 15.2 should not place unnecessary demands on the participants. The use of telephone, video or online conferencing works well, particularly with multiple participants in different locations.

Clause 15.4 goes on to say that the Project Manager revises the Early Warning Register after the early warning meeting and issues a copy to the Contractor within one week of the meeting. Whilst Clause 15.3 mentions removing risks from the Early Warning Register the author recommends retaining them, perhaps in an appendix or struck through, for future consultation.

1.8 Contractor's proposals

Clause 16 is a new provision, where the Contractor may propose to the Project Manager that the Scope is changed to reduce the amount the Client pays to the Contractor. This could also be considered as part of a value engineering or a risk management process.

An example could be that a material specified within the Scope could be substituted with another, less expensive material that is equally effective, or a specified methodology could be substituted with a more rational, and again less expensive methodology, for example the use of mobile access platforms rather than a full static scaffold.

Within four weeks of making the proposal, the Project Manager may:

- accept the proposal and issue an instruction changing the Scope;
- inform the Contractor that the Client is considering the proposal and instruct the Contractor to submit a quotation; or
- inform the Contractor that the proposal is not accepted.

1.9 Adding to the Working Areas

The Contractor may also submit a proposal for adding to the Working Areas to the Project Manager for acceptance.

The implications of adding to the Working Areas depend on the Main Option chosen:

Option A

The Short Schedule of Cost Components refers to the cost of resources used within the Working Areas, so these resources if also used within the extended working areas would be included as cost rather than within the fee percentage when assessing compensation events.

Options C, D and E

The Schedule of Cost Components refers to the cost of resources used within the Working Areas, so these resources if also used within the extended working areas would be included as cost rather than within the fee percentage for payments, and also when assessing compensation events.

A reason for the Project Manager not accepting is that the proposed area is:

- not necessary for Providing the Works;
- used for work not in the contract.

2 Contractor's main responsibilities

2.1 Introduction

The Contractor's main responsibilities are covered within the ECC by Clauses 20.1 to 29.2.

This section of the contract includes Contractor's design, people, working with the Client and Others, subcontracting, assignment and disclosure. Due to the wide range of the Contractor's responsibilities some are covered within this chapter, some in other chapters.

2.2 Contractor's design

The ECC provides for all, or part of the design to be carried out by the Contractor, recognising that it is commonplace for clients to assign design to the Contractor.

This may be in the form of parts of the works to be designed by the Contractor, or a full design and build arrangement.

Note that there is no separate NEC4 Design and Build Contract; design is assigned to the Contractor through the Scope. (N.B. There is a separate NEC4 Design Build and Operate Contract.)

Whilst many believe design and build to be a recent innovation, it is a procurement method which has been in use for hundreds of years but has re-emerged as a commonly used form of procurement in the past 25 years. Before the emergence of architecture as a profession, clients used to procure buildings by a process of design and build. It was the separation of responsibility for construction from responsibility for design that led to the emergence of so-called "traditional contracting" in the nineteenth century.

Examples of successfully completed design and build projects are too numerous to mention. They range from housing, through industrial and commercial projects, to major refinery and sporting complexes. It is clear that this procurement method has a very wide applicability.

Advantages of Design and Build

- single-point responsibility, the Client has only one party (the Contractor) to deal with rather than a contractor and a number of consultants, so the traditional design/construction interface and the risks associated with it are transferred to the contractor. Whether a defect is a design fault or a workmanship fault, it is the responsibility of the Contractor to correct that defect.

• inherent buildability is achieved because the Contractor designs the project so that it can be built in a practical, rational and economical way. The Contractor can utilise its experience and expertise in using preferred methods of construction to minimise time and cost. Separation of the "design" and "build" elements and lack of co-ordination between designer and builder can lead to designs which are less buildable and in turn more expensive.

No Contractor's Design

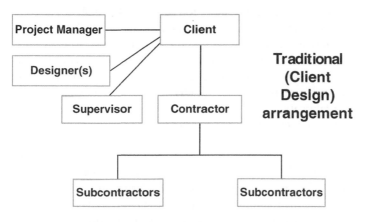

Full Contractor's Design

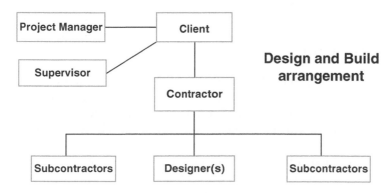

Figure 2.1 Client's or Contractor's design arrangements

- price certainty is obtained before construction starts, provided the Client's requirements are clearly and adequately specified and changes are not introduced. The management of the design risk by the Contractor can result in greater certainty of the time, cost, and project objectives being met.
- it is good for "non expert" clients who do not have the knowledge or experience to be able to select, appoint and enter into contracts with a group of consultants of varying disciplines. It engages the Contractor as a single point of responsibility to provide that expertise.
- there is a reduction in procurement time from conception to start on site, for example, there is no requirement for the Client to complete the design and prepare bills of quantities etc. prior to tendering and appointing the Contractor. There is also potentially a reduced period on site, as design and construction can overlap.

Disadvantages of Design and Build

- the Client is required to commit itself before the detailed designs are completed. Clients often find this difficult to do so, as the Client has reduced control of design and quality as these are passed to the Contractor at an early date. However, this can be overcome if outline designs are prepared at tender stage.
- the Contractor designs and builds to a price commitment, therefore where there is a conflict between aesthetic quality and ease of fabrication, the requirements for fabrication will probably override.
- there is no design overview unless separate consultants are appointed by the client for this purpose.
- clients often maintain that complex or prestigious construction projects cannot be procured using design and build, however the problem is with the adequacy of the Client's requirements rather than design and build itself.
- bids are difficult to compare since each design, programme and cost will vary.
- whilst the procurement period may be shortened, the Contractor must be given more time to prepare its tender.
- it is expensive for the Contractor to prepare its tender as it needs to consult with designers as well as to price the tender.
- the Client has reduced control of change costs, and client changes to the works can be expensive and are also more difficult to assess as there is usually a brief contract sum analysis rather than a traditional bill of quantities.

If the Contractor is to design or all part of the works, the Client must create a clear and unambiguous set of requirements within the Scope, clearly defining what it requires the tendering contractors to price and eventually to carry out. The Contractor responds with its proposals, which will include production as well as design work.

The Contractor's design input varies from one contract to another, ranging from the mere development of a detailed design which was provided by the Client, to a full design process including proposals, sketch schemes and production information.

There will often be some negotiation between the Client and the Contractor, with the aim being to settle on an agreed set of requirements. These will include the contract price, as well as the manner in which it has been calculated.

The most important advantage with design and build is that the Contractor is responsible for everything, this "single-point" responsibility being very attractive to clients, particularly those who may not be interested or able to distinguish the difference between a design fault and a workmanship fault. This single-point responsibility also means that the Contractor is not relying on other firms (e.g. architects employed by the client) for the execution of design, or for the supply of information.

The design and build process increases the opportunities for recognising and harnessing the Contractor's experience during the design stages of the project.

Many of the developments in procurement processes, such as ECI (Early Contractor Involvement) and much of the work in the field of study of buildability and value engineering, have had this purpose in mind. The benefits of the integration of designers and builders are more rational, economic projects in additional to better consideration of life cycle issues.

One reported disadvantage of design and build is where there is a conflict between aesthetic quality and ease of construction; as the contractor is also designing the project, the requirements for construction will override.

A further criticism has been that a design and build contractor will put in the minimum effort with regard to the design in an effort to secure the contract. These two criticisms suggest strongly that quality, particularly design quality, will suffer under this procurement process. However, this is not a valid criticism of design and build as a procurement form, but rather of the people who advise clients and prepare the Client's requirements and tender documents.

When selecting a contractor, the Client may approach design and build contractors who will prepare the design and submit their tenders based on a "shopping list of requirements", or in many cases will employ its own designers to carry out initial designs of major elements of the works, which would then form the basis of the Client's requirements, with the designers transferred to the Contractor by a novation agreement.

Design management

Good design and its management, whichever party is responsible for it, is an underrated but key component in the success or failure of construction projects.

It is imperative that the client understands what it wants and is able to clearly communicate that to the designer in the form of a clearly structured brief. The designer can then convert that brief into a practical and economic solution delivered meeting quality standards, and within budget and available time constraints.

Whilst quality and financial constraints tend to be uppermost in clients' minds it is critical that sufficient time is allowed for the design to be developed and alternatives considered, then the often time-consuming process of submission and awaiting regulatory approval, before making final decisions.

In the European Union consideration must also be given when designing projects in the public domain to time scales to comply with OJEU regulations.

Key points in effective design management:

- clients should always be clear as to their objectives, priorities and operational requirements including the needs of all stakeholders before appointing designers and ensure that these are clearly stated in their brief. The initial stages are the most important part of any project, and a well-informed client is the critical foundation to the successful delivery of the project within the budget, agreed timescale and to the required standards.
- allocate sufficient time and resources to establish the client's design quality aspirations ensuring that design time is not reduced in order to recover time lost in the programme for other reasons – good design takes time. Time allocation should also include the need for external regulatory approvals.
- appoint someone responsible for taking the lead on design issues and co-ordinating the work of the various parties.
- continue to communicate design requirements throughout the design, procurement and construction stages.
- good design should consider the most economical solution both in terms of initial cost, but also life cycle cost including future maintenance, therefore value engineering and continuous design review are integral to the process.
- good design should consider not only aesthetic and structural features, but also functionality and efficiency, build quality, accessibility, sustainability, whole life costing and flexibility in use.
- clients should always be clear as to what funding they have available for the design process and for what is to be designed. That funding should always be realistic.
- a candid and honest, but formal post-contract audit should always be carried out to ascertain the successes and failures of the project from which lessons can be learned and carried through to the next one.

Selection of the design team

There are several methods of selecting the appropriate design team for a particular project including, particularly for a high profile public project, design competitions, which primarily select specific design ideas or outline designs for a project, rather than the individual design team members.

The key issue with design competitions is that the client must have a clear vision of what it requires, in order that design teams have clear criteria from which to create their designs.

Design competitions may be appropriate where there is either a unique problem that will benefit from a wide range of design approaches being explored (along with likely considerable public interest – which may be the case on a major new public building) or where the competition promoter wishes to encourage the development of new talent.

Alternatively one can focus on the individual designers and their skills and reputation in designing similar projects. Again it is vital that not only aesthetic qualities are considered, but also the design team's ability to perform within the time constraints and budget.

All public sector appointments, irrespective of the client's preferred nature of competition or reference to any other guidance on design competitions, must be consistent with EC procurement rules in terms of process and outcome.

Whichever process is used for selecting the design team there must be clear and objective criteria on which to select and appoint the successful designers and this should consider:

- the balance between quality, price and time related issues.
- the technical ability of the design team to deliver the project on time and within budget, as well as to the required standards of quality.
- the financial standing of the designers including insurance requirements, etc. The financial failure of a designer can be catastrophic to the successful completion of the project.

The NEC4 is unique amongst construction contract families in that, whilst most other contracts provide for portions or all of the design to be carried out by the contractor, they all use a separate contract for design and build where the contractor has full responsibility for design.

In reality, there is no need for a separate design and build contract as within the ECC the clauses which cover early warnings, programme, payment, change management etc. are the same whether or not the Contractor has design responsibility.

The "default position" within the contract is that all design is carried out by the Client, but any design to be carried out by the Contractor is assigned to him within the Scope, and Clauses 21–23 cover design obligations and submission and acceptance of the Contractor's design proposals.

The most important advantage with design and build is that the Contractor is wholly responsible for the design. This "single-point" responsibility is very attractive to clients, particularly those who do not have the knowledge or experience to appoint design consultants, or who may not be interested or able to distinguish the difference between a design fault and a workmanship fault.

This single-point responsibility also means that the Contractor is not relying on other firms (e.g. architects) for the execution of design or for the supply of information.

The design and build process increases the opportunities for harnessing the benefit of the Contractor's experience during the design stages of the project. Many of the developments in procurement processes, and much of the work in the field of study known as "buildability", have had this purpose in mind. The benefits of the integration of designers and builders are more economic buildings as well as a more economic and effective production process.

The technical complexity of the client's process may lead to solutions which are more akin to engineering than to building. While this does not mean that design and build is unsuitable, it is likely that in such circumstances the use of other forms of procurement will be favoured.

2.3 Contractor's design provisions within the ECC

The Scope

When preparing the Scope which includes contractor's design, it is important that the right balance of information should be considered. If it is too prescriptive it will lead to all the tendering contractors submitting similar designs and prices. If the Scope is not detailed enough, then none of the tendering contractors will produce a design which is as the client intended.

As an example, the client wishes to invite tenders for the design and build of 52 houses. The client has certain essential requirements, but wishes to attract some innovation in that contractors will interpret the shape, size and appearance of the houses in different ways subject to the approval of external bodies such as the local authority planners and other legislative bodies, etc.

The Scope should therefore comprise the following groups:

Essentials – requirements that the client must have

> For example, if the contractor is to design and build houses then the number of houses and the mix of types of houses for example number of bedrooms, should be clearly stated. Any other essentials should also be clearly stated, otherwise a tenderer may not include them.

Preferences – requirements that the client would prefer to have

> For example, the client may have a preference for a certain heating system, but the contractor may offer an alternative which is more efficient, cheaper to run, etc.

> The client may also state within its preferences that "equal and approved" may be acceptable subject to the contractor proving that it truly is "equal" and also that it is prepared to approve it. As part of this proof the contractor should consider issues such as life cycle costing. For example, if the contractor can prove that the alternative heating system is equally efficient and easy to operate, but in the longer term the alternative requires a higher level of maintenance then the client will need to be made aware of this requirement. It is suggested that the preferences should not applicable to critical aspects of the project.

Innovations – something the client may not have thought of that the contractor believes could win the tender!

> This is open to the contractor to provide as added value its particular bid. For instance, one contractor may offer a children's playground, or a particularly "green" design/construction over and above the requirements of the Scope and/ or the relevant legislation. If a contractor feels it can exceed the Scope it should submit its proposals which comply with the document (a compliant bid) but list separately how it could exceed it and any time and/or price implications.

With regard to the contractor's proposals within its tender, it will wish to demonstrate that its bid is compliant with the Scope, though it is important that it is not prescriptive about exactly what it will provide, as its design may evolve and develop as the construction progress. It may decide on new materials, or it may wish to use some monies to offset any problems that may arise.

As a more complex example, the client wishes to procure a new office block on a full design and build basis. Typically it should consider the following:

Quality

What are its overall requirements?

1   The size and use of the building. If no specific size of building is known, details of exactly what the client intends to use the proposed building for and expected number of occupants.
2   Open plan or cellular offices.
3   Any specific requirements, for example canteen facilities, gym, etc. Number of floors, clear span, or sub divisions.
4   Need for future flexibility and expansion.
5   Specific heating requirements, lighting levels for various parts of the building, air conditioning requirements. Running costs and energy conservation requirements.
6   Car parking facilities.
7   Maintenance, running costs and life cycle cost factors. How long is the building expected to last?

Time

When will the client need the building?

1   Is the starting date and/or completion date critical to the client?
2   Consider pre-tender activities such as planning consents, preparation of tender documents and also other operations which need to be carried out before the contractor can be appointed and start on site.

Cost

How much will be paid?

1   What is the budget for the building? How tight is that budget? Is there a cost plan and expected cash flow projections for the building? How long will it take to prepare one?

2   Is the project being wholly funded by the client or is partial funding coming from another source, e.g. grant aid? How much influence do the funders have in terms of the design and construction of the building?

From these initial considerations the documentation to be prepared for tender can be established:

1   What drawings are to be included within the Scope? Broad sketch plans, showing size and general layout only? Or detailed design drawings?
2   Site survey and particularly ground exploration – by client or contractor? Who is responsible for accuracy?
3   Who will apply for outline and later detailed planning consents? Possible delays? Revised applications due to changes in design suggested by contractor?
4   Performance Specification? Or specification of work and materials? Outline Spec or detailed at this stage?
5   Any major products, specialist or supply items over which the client wishes to exercise some choice or control? Nomination is difficult with design build and creates complications.
6   Client to provide an outline time programme, or left to the contractor to submit with the tender?
7   Staged Completions? Clients fitting out, timing, split responsibilities and insurances.
8   Constraints – access, space, working hours, sequence of working? Client's or statutory requirements, if known.
9   Which Main and Secondary Options are to be used?
10 Design and Construction Liability – any special insurance requirements or performance bonds? Design warranties?

Other documentation

From the above it can be readily seen that it is vital that the client defines how much information and detail is required in the returned tenders, and it is crystallised within the Scope.

For example, a full set of drawings and comprehensive specification would not normally be expected at tender stage and the scope and limitation of these documents must be defined.

The following must be clearly stated:

(i)   Details of the Site and its boundaries.
(ii)   Details of the accommodation requirements.
(iii) Purpose for which the building is to be used.
(iv)   Statements of functional requirements, e.g. kind and number of buildings, density and mix of accommodation, schematic layouts and/or drawings.
(v)   Specific requirements as to finishes and any other elements.
(vi)   Requirements for "as built" drawings.

(vii)   Functions to be carried out by the Project Manager, and any other members of the client team.
(viii) Statement of any planning restraints, e.g. restrictive covenants, and any permissions already granted.
(ix)   Access restrictions.
(x)   Availability of public utilities.
(xi)   Various other information to be included in the contract.
(xii) Any other matters likely to affect the preparation of the contractor's tender.

In the Engineering and Construction Contract "the Contractor Provides the Works in accordance with the Scope" (Clause 20.1) and "the Contractor designs the parts of the works which the Scope states the Contractor is to design" (Clause 21.1).

It is clear, then, that the Scope should clearly show what is to be designed by the contractor and this could consist of detailed specifications or some form of performance criteria and certain warranties.

When clients are preparing the Scope and considering any of the works to be partially designed by the contractor the following should be reviewed:

(i)   What portions of the work would be better designed by the contractor? Does the contractor have particular expertise with that element of the works?
(ii)   Who is responsible for the interface between the part that is designed by the contractor and the part that is designed by the client's designers?
(iii) What is the overall contribution and co-ordination by the lead/principal designer? The responsibility for the overall design is the client's, the contractor is only responsible for the part that it has designed.

2.4 Acceptance of design

The Contractor is required to submit the particulars of its design as the Scope requires to the Project Manager for acceptance (Clause 21.2).

A reason for not accepting the Contractor's design is that it does not comply with either the Scope or the applicable law.

The particulars of the design in terms of drawings, specification and other details must therefore be sufficient for the Project Manager to make the decision as to whether the particulars comply.

The Scope may stipulate whether parts of the design may be submitted in parts, and also how long the Project Manager requires accepting the design. Note that in the absence of a stated time scale for acceptance of design, the "period for reply" will apply.

Many Project Managers are concerned that they have to "approve" the Contractor's design and therefore they are concerned as to whether they would be qualified, experienced and also insured to be able to do so. Note the use of the word "acceptance" as distinct from "approval". Acceptance denotes compliance with the contract, it does not validate that the design will work, that it will be approved by regulating authorities, or that it will fulfil all the obligations which the contract and the law impose, therefore performance requirements such as structural strength, insulation qualities, etc. do not need to be considered.

Whether the design particulars comply with the law is purely an observation on behalf of the Project Manager. Ultimately, it is the Contractor's responsibility to ensure that its design works and also complies with the applicable law, and any acceptance or non-acceptance by the Project Manager, will not absolve the Contractor of that responsibility in any way.

Note that under Clause 14.1, the Project Manager's acceptance does not change the Contractor's responsibility to Provide the Works or its liability for its design. This is an important aspect as, if the Project Manager accepts the Contractor's design but following acceptance there are problems with the proposed design, for example, in meeting the appropriate legislation or the requirements of external regulatory bodies, this is the Contractor's liability.

The Contractor cannot proceed with the relevant work until the Project Manager has accepted its design.

Under Clause 27.1, the Contractor obtains approval of its design from Others where necessary. This will include planning and inspection authorities, and other third party regulatory bodies.

Note also, Clause 60.1(1) second bullet point:

> The Project Manager gives an instruction changing the Scope except a change to the Scope provided by the Contractor for its design which is made – in order to comply with the Scope provided by the Client.

which clearly refers to the fact that the Client's Scope prevails over the Contractor's proposals.

This clause gives precedence to the Scope in Part 1 of the Contract Data over the Scope in Part 2 of the Contract Data. Thus the Contractor should ensure that the Scope it prepares and submits with its tender as Part 2 of the Contract Data, complies with the requirements of the Scope in Part 1 of the Contract Data.

2.5 Defects in Contractor's design

Defects are defined in Clause 11.2(6) as:

- a part of the works which is not in accordance with the Scope

or

- a part of the works designed by the Contractor which is not in accordance with the applicable law or the Contractor's design which the Project Manager has accepted.

2.6 Client's Scope v Contractor's proposals

The ECC is unlike most other construction contracts in that it does not provide a priority or hierarchy of documents, but one can "read into" the various clauses as to which takes precedence in the event of an inconsistency.

A question to then be considered is, if the Contractor has full design responsibility and there is an inconsistency between the Scope provided by the Client and the proposals provided by the Contractor, which takes precedence?

The answer can be found within Clauses 11.2(6) and 60.1(1).

Defects are defined in Clause 11.2(5) as:

- a part of the works which is not in accordance with the Scope

or

- a part of the works designed by the Contractor which is not in accordance with the applicable law or the Contractor's design which the Project Manager has accepted.

Clause 60.1(1) states:

The Project Manager gives an instruction changing the Scope except

- a change made in order to accept a Defect or
- a change to the Scope provided by the Contractor for its design which is made either at its request or to comply with other Scope provided by the Client.

2.7 Use of the Contractor's design

Under Clause 22.1, the Client may use and copy the Contractor's design for any purpose connected with construction, use, alteration or demolition of the works unless otherwise stated in the Scope and for other purposes as stated in the Scope.

The Contractor is also required to obtain similar rights from Subcontractors for the Client to use the Subcontractor's design.

If the Client wishes to own the rights to any design material prepared by the Contractor as part of the contract, then the Client should include an appropriate Z clause to provide for it.

2.8 Advising on the practical implications of the design of the works

Under Main options C, D, E and F, the Contractor has additional obligations in respect of design in that, whether it has design responsibility or not, it is required to advise the Project Manager on the practical implications of the design of the works.

2.9 Fitness for purpose v reasonable skill and care

In order to consider the design responsibilities of the parties, we should consider the principles of "fitness for purpose" and "reasonable skill and care".

"Fitness for purpose" means producing a building, part of a building, or a product, fit for its intended purpose. This is an absolute duty independent of negligence. In the absence of any express terms to the contrary, a Contractor or Subcontractor who has design responsibility will be required to design "fit for purpose".

"Reasonable skill and care" means performing to the level of an ordinary competent person exercising a particular skill. The criterion is that it is not possible or reasonable to judge someone's professional ability purely by results, e.g. a lawyer whose client is found guilty is not necessarily a poor lawyer, a surgeon whose patient dies is not necessarily a poor surgeon. The question to ask is "did that professional apply the correct level of skill and care in carrying out their duties?"

If one then applies the same criterion to a designer, this is normally achieved by the designer following accepted practice and complying with statutory requirements, codes of practice etc., but if a new construction technique is involved the duty can be discharged by taking advice from consultants and warning the Client of any risks involved. A further qualification on the designer's liability is the "state of the art" defence, meaning that a designer is only expected to design in conformity with the accepted standards of the time.

In the absence of any express terms to the contrary, a designer will normally be required to design using "reasonable skill and care".

Design and build contracts can impose a higher standard than reasonable skill and care, i.e. fitness for purpose. This obligation resembles a seller's duty to supply goods which are fit for their intended purpose.

The "default" position within the ECC is that any design carried out by the Contractor must in all respects be fit for purpose. This means producing a building, part of a building, or a product, fit for its intended purpose. This is an absolute duty independent of negligence.

In the absence of any express terms to the contrary, a construction contract which includes design, requires the Contractor to design "fit for purpose".

The JCT 2016 Design and Build Contract states that "the Contractor shall in respect of any inadequacy in such design have the like liability to the Client, whether under statute or otherwise, as would an architect or, as the case may be, other appropriate professional designer holding itself out as competent to take on work for such design who, acting independently under a separate contract with the Client, has supplied such design for or in connection with works to be carried out and completed by a building contractor who is not the supplier of the design" – the JCT version of "reasonable skill and care"!

Under Option X15, the implied "fitness for purpose" obligation can be changed to a "reasonable skill and care" requirement which is effectively a relaxation of the default position.

"Reasonable skill and care" means designing to the level of an ordinary competent person exercising a particular skill. This is normally achieved by the designer following accepted practice and complying with standards, codes of practice, etc.

Note the "state of the art" defence that could be applied in the case of a new type of material or a new and as yet untried construction technique in which case the duty can be discharged by taking advice from consultants and warning the Client of any risks involved.

In the absence of any express terms to the contrary, a contract for design requires the designer to design using "reasonable skill and care".

The Contractor is not liable for Defects in the works due to its design so far as it proves that it has used reasonable skill and care to ensure that its design complied with the Scope.

If the Contractor corrects a Defect for which it is not liable under this contract it is a compensation event.

2.10 Design of Equipment

The Contractor may be required to carry out design of Equipment, which could include temporary work such as a temporary footbridge or a specialist scaffold structure, and also what would in other contracts be defined as "Plant".

Under Clause 23.1, the Contractor submits particulars of the design of an item of Equipment to the Project Manager for acceptance, but only if the Project Manager instructs him to. This is as distinct from design of the "works" for which the Contractor is required to submit particulars of its design.

A reason for not accepting the design of the item of Equipment is that the design of the item of Equipment will not allow the Contractor to provide the works in accordance with the Scope, the Contractor's design which the Project Manager has accepted, or the applicable law.

Again, as with design of the permanent works, whether the design particulars comply with the law is purely an observation on behalf of the Project Manager. Ultimately, it is the Contractor's responsibility to ensure its design works and also complies with the applicable law, and any acceptance by the Project Manager will not absolve the Contractor of that responsibility in any way.

2.11 Intellectual Property Rights

The question of Intellectual Property Rights (IPR) which includes copyright, patents and trademarks needs to be considered carefully where the Contractor is designing all, or part of the works.

The design and construction of a building or other structure is unique and unless the contract states otherwise, even though the owner pays for the design work, it does not automatically own the copyright in the design documents; the designer retains ownership and control over the design and the various documents, based on copyright law and the terms of agreement between the client and the designer.

This copyright issue can be very complex as it only applies to that which was originally conceived by the designer and therefore much of what is included in the design may not be subject to copyright, for example the layout of kitchen units or sanitary appliances within a bathroom.

Copyright extends to copying the design, photocopying and distributing drawings and other documents, etc.

In the event that copyright is breached, the owner of the copyright may obtain an injunction preventing further breaches and if necessary may seek impounding of materials, and also damages, loss of profits, and legal fees. If a building design copyright has been breached the client may be prevented from commencing or continuing with the building works.

2.12 Collateral warranties

A collateral warranty is an agreement associated with another contract, where collateral means "additional but subordinate" or "running side by side", with the collateral warranty being entered into by a party to the primary contract, for example a Contractor or a Subcontractor, and a third party who is not a party to the primary contract but who has an interest in the construction project.

Collateral warranties are a relatively recent feature of contracts, coming into use in the past 20 years following some notable court cases in which it was judged that it was not possible to recover damages for negligence in relation to defects in construction projects as it was held to be economic loss, which is not recoverable in tort.

Consequently, such claims had to be brought as claims for breach of contract, where economic loss is recoverable. It therefore became necessary to establish a contract between parties involved in a construction project, such as designers or subcontractors and parties such as funders, present and future purchasers, who are not parties to the primary contracts but who have an interest in the project, hence the "collateral warranty".

The NEC4 contracts do not provide for collateral warranties, though it is worth some commentary on their use and how they could be included specifically within the ECC.

Collateral warranties are different to traditional contracts, in that the two parties to the warranty do not have any direct commercial relationship with each other. They are useful for creating ties between third parties, but the main disadvantage of collateral warranties is the expense of providing them, particularly where there are many interested third parties such as with housing and commercial properties. When one also considers that parties can change their names or business status during the life of a warranty then even a medium-sized project can involve 30 or more warranties.

Different parties will also have different requirements for collateral warranties, therefore the precise terms of the collateral warranty agreement need to be considered in each case carefully so that they are drafted to provide the required protection.

Collateral warranties are executed as a "deed" also referred to as "under seal", which differs from a simple contract in two respects:

- Firstly, in English law the time after a breach of contract has occurred within which one party can sue another is 12 years, whereas the time for a simple contract that is executed "under hand" is 6 years.
- Secondly, whereas under English law consideration is needed for a contract to be effective, this is not the case with a deed. For a company to execute a deed effectively, the document must be signed by two directors or a director and the company secretary, unless the Articles of Association contain some other requirement.

The basic ingredients of a collateral warranty are usually:

(i) Definitions

There will be a brief description of the works and the collateral warranty. The "contract" is the contract to which the collateral warranty agreement is collateral and the date should be specified. Since the obligations of the Contractor to the beneficiary parallel the obligations of the contractor to the developer under the contract for the works, the beneficiary should always ask to see a copy of the contract. If, for example, there are restrictions on the Contractor's liability in the principal contract, that might affect the rights of the beneficiary under the collateral warranty.

A copy of the principal contract may be attached to the collateral warranty agreement.

(ii) The warranty

The Contractor gives warranties to the beneficiary that it has exercised and will continue to exercise reasonable skill and care in its obligations to the client under the principal contract including compliance with any design obligations, performance requirements and correction of defects.

The warranty should contain a statement that the warrantor shall have no greater liability to the beneficiary than they would have under the primary contract.

The Contractor also does not have any liability to the beneficiary if it fails to complete the works under the contract on time, as that will usually be covered by delay damages under the principal contract.

There should be a statement that the warrantor shall have no liability under warranty in any proceedings commenced more than 12 years after the date of completion of the works.

(iii) Naming of the parties

The named parties can be, for example, the Contractor and the end user, a Subcontractor and the client, a Consultant and the client.

The party that gives the warranty is known as the warrantor, the other party is known as the beneficiary.

(iv) Prohibited materials

    The Contractor may be required to give a warranty that it will not specify or use materials which are known to be prohibited or deleterious to health and safety.

    In addition, the use of deleterious materials may contravene relevant legislation or regulations as well as being expressly prohibited in the contract between the developer and the contractor.

(v) Copyright

    Normally, the beneficiary under a collateral warranty agreement will have a right to use designs and other documents prepared by the Contractor but only in connection with the project for which those design documents are prepared, copyright to these documents remaining with the Contractor.

(vi) Insurances

    There is normally a requirement for the warrantor to maintain insurances including professional indemnity (PI) insurance where applicable for a stated minimum level of cover and for a period of 12 years after completion of the works.

(vii) Assignment

    There is normally a provision allowing the beneficiary up to two assignments of the warranty.

(viii) Step in rights

    Step in rights can allow a funder to take over the client's role and become the client under the primary contract. This is particularly important where the client is unable to pay the Contractor, the Contractor having to notify the beneficiary before it terminates the contract, this allowing the beneficiary to "step in" and take the client's place, with an obligation to pay the Contractor any money that is outstanding. The beneficiary does not have any obligation to step in, only the right to do so. As the effect of this clause is to vary the principal contract it is important that, when the collateral warranty contains such a clause, the client should be a party to that document.

(ix) Law and jurisdiction

    It is important to specify the law of the contract and the warranty and the jurisdiction of the courts will have jurisdiction.

(x) Dealing with planning permission

    The responsibility for obtaining planning permission can lie with the client or can be assigned to the Contractor.

In some cases, in a design and build contract, the Contractor may be responsible for gaining outline planning for the project, however this should be a two-stage process so that if outline planning permission is not obtained the second stage would not proceed. It is more usual in design and build for the client to gain outline planning permission and the Contractor to be responsible for gaining full and detailed planning consent.

2.13 Novation of designers

Whilst the principle of design and build agreements is that the Client prepares its requirements and sends them to the tendering contractors and they prepare their proposals to match the Client's requirements, in reality in nearly half of design and build contracts, the Client has already prepared initial designs before tenders are invited from design and build contractors. Outline planning permission and sometimes detailed permission will have been obtained for the scheme before the Contractor is appointed.

Each Contractor then tenders on the basis that the client's design team will be novated to the successful Contractor who will then be responsible for appointing the team and completing the design under a new agreement.

This process is often referred to as "novation", which means "replace" or "substitute" and is a mechanism where one party transfers all its obligations and benefits under a contract to a third party. The third party effectively replaces the original party as a party to that contract, so the Contractor is in the same position as if it had been the client from the commencement of the original contract. Many prefer to use the term "consultant switch" where the consultant "switches over" to work for the Contractor under different terms as a more accurate definition.

This approach allows the client and its advisers time to develop their thoughts and requirements, consider planning consent issues, then when the design is fairly well advanced, the designers can then be passed to the successful design and build Contractor.

The NEC4 contracts do not have pro forma novation agreements, but it is critical that the wording of these agreements be carefully considered as there are many badly drafted agreements in existence! Novation can be by a signed agreement or by deed. As with all contracts, there must be consideration, which is usually assumed to be the discharge of the original contract and the original parties' contractual obligations to each other. If the consideration is unclear, or there is none, you should use a novation agreement which is drawn as a deed.

In many cases the agreement states briefly and very badly that from the date of the execution of the novation agreement, the Contractor will take the place of the client as it had employed the designers from the beginning, the document stating that in place of the word "Client" one should read "Contractor", however the issues of design liability, inspection, guarantees and warranties may not have been considered and drafted into the agreement in sufficient detail and referring to the correct parties.

In practice, the best way is to have an agreement drafted between the client and the Designer covering the pre-novation period, and a totally separate contract drafted between the Contractor and the Designer for the post-novation period.

When a contract is novated, the other (original) contracting party must be left in the same position as it was prior to the novation being made. Essentially, a novation requires the agreement of all three parties.

Often clients engage in what can only be described as "partial novation", by which design responsibility is passed to the Contractor, but the designer continues to work for the client in providing an ongoing advice service including inspection of the works and certification.

How to make novation work

- The contract must do more than simply stating that, as far as the consultant is concerned, the contractor is the client.
- The consultant's appointment must be novated to the contractor before detailed design.
- Avoid partial novation – the consultant may face a conflict of interest if it is required to act for more than one party.

As this book is intended as a guide for NEC4 practitioners worldwide it is not intended to examine specific legal cases as they may not apply on an international basis, but there is certainly case law available in terms of design errors in the pre-novation phase being carried through as the liability of the Contractor in the post-novation stage, and also the accuracy and validity of site investigations and the Client's and the Contractor's obligations in terms of checking and taking ownership of such information.

2.14 People

Key people

The Contractor is required to name each "key person" within Contract Data Part 2, including their name, job, responsibilities, qualifications and experience.

In Providing the Works he is required to employ the same people, or a replacement that has been accepted by the Project Manager, including the requirement to submit details of the replacement's name, job, responsibilities, qualifications and experience. The Project Manager may refuse to accept the replacement if their qualifications and experience are not as good as the person they are replacing.

Removing a person

Under Clause 24.2, the Project Manager may instruct the Contractor to remove a person employed by the Contractor. The Project Manager must state his reasons for giving such an instruction.

Many practitioners have queried what sort of reasons a Project Manager could give for removing a person, typical reasons usually being:

- the person has infringed a health and safety requirement, possibly creating a danger to himself and to others;
- the person is not competent to carry out his duties;
- the person has behaved aggressively towards the Client, the Project Manager or another party.

Clearly the Project Manager cannot give a reason that is illegal under the law, for example instructing the Contractor to remove a person purely on the basis of their race or gender.

An interesting example used once was that the Project Manager instructed the Contractor to remove a person as he did not like him, which, unless it is considered as a breach of the obligation under Clause 10.2 to act in a spirit of mutual trust and co-operation can only be considered as a valid reason!

2.15 Working with the Client and Others

The ECC uses the term "Others" to define parties who are "not the Client, the Contractor, the Adjudicator or any employee, Subcontractor or supplier of the Contractor", which could include other contractors, utilities companies, third party regulators, etc.

Clause 25.1 requires the Contractor to cooperate with Others in obtaining and providing information which they need in connection with the services. Under Clause 25.1, the Contractor is also required to obtain approval from Others where necessary.

The Contractor may also have to share the Working Areas with Others as stated in the Scope. The Scope may also require the Contractor to coordinate the work with others including holding or attending meetings with them, and notifying the Client before the meetings in case the Client wishes to attend them.

2.16 The Client's obligations

Clause 25.2 covers the Client's obligations to provide services and other things which the contract requires him to provide in accordance with the Scope.

The use of the terms "thing" and "things" is a little curious, and usually applies to something that cannot be adequately described! However, it applies to facilities, services, etc. which the Client provides, and should be fully described in the Scope and also specifically referred to in Contract Data Part 1 which identifies what access is to be given, and when.

It is critical that the parties understand what is to be provided by the Client, and it is also critical that if there is a charge for the availability and use of those things, it should be clearly detailed within the contract, including how the charge will be measured and when the Contractor has to pay.

The Project Manager may give an instruction to the Contractor which changes the Scope or a Key Date; however, although it is not expressly stated within the contract, the Project Manager may not give an instruction which could require the Contractor to act in a way which was outside his professional code of conduct.

Contractors (and Consultants) are normally required to adhere to Codes of Conduct which include a duty to behave ethically, and are normally set by the professional institution or trade bodies of which they are members.

Examples of Codes of Conduct normally include obligations to:

(i) discharge professional duties with integrity and in a fair and unbiased matter, to uphold and enhance the standing and reputation of the relevant institution;

(ii) only undertake work that they are professionally and technically competent to do and in carrying out that work to adhere to good practice and to demonstrate a level of competence consistent with a professional of their standing. This may also be coupled with the requirement to maintain insurances and indemnify their clients against risks;

(iii) have full regard for the public interest, particularly in relation to matters of health and safety, but also to others' business activities, and to the wellbeing of future generations;

(iv) show due regard for the environment and for the sustainable management of natural resources;

(v) develop their professional knowledge, skills and competence on a continuing basis and give all reasonable assistance to further the education, training and continuing professional development of others;

(vi) not improperly offer or accept gifts or favours which would be interpreted as exerting an influence to obtain preferential treatment.

Examples of a Project Manager giving an instruction which could require the Contractor to act outside his professional code of conduct would be instructing him to do something which was outside his area of expertise, which could cause harm to the Contractor or to others, or to do something which would infringe another's copyright, or put him in a position where there would be a conflict of interest.

3 Time

3.1 Introduction

Time is covered within the ECC by Clauses 30.1 to 36.3, and also by Clause 31.4 within Main Option A.

3.2 General principles of planning and programme management

A successful construction project is considered by clients and consultants to be one which has been completed on time, to the required quality and also within budget, and by contractors and subcontractors as one which has been completed on time, to the required quality and making a profit. Clearly, apart from their different perspectives on financial outcome, the objectives of both contracting parties are well matched.

When considering completion on time, various surveys carried out in recent years have shown that only about 33 per cent of contracts are completed within the contract period, one fifth overrun by 40 per cent and a significant number overrun by more than 80 per cent!

There are many factors which can cause delays:

(i) *Delays/errors in design, often exacerbated by poor communications between design and construction teams.*
(ii) *Inadequate site investigations, clients often believing that contractors should be responsible for assessing ground conditions based on poor and inadequate information provided by them.*
(iii) *Inability to give timely access or possession.*
(iv) *Non-availability of resources, by all parties.*
(v) *Poor performance of parties.*
(vi) *Client changes, compounded by the number or timing of these instructions.*
(vii) *Weather and other seasonal problems.*
(viii) *Insolvency of a party.*
(ix) *Unrealistic construction period set by client.*

Clearly matters need to be improved by the parties and between the parties, but many of these factors can be avoided or at least reduced by the parties engaging in proper and effective planning and programming of the works aligned to effective communication between the parties.

Planning and programming

The terms "planning" and "programming" are synonymous, however they are a little different.

(i) What is planning?

Planning may be defined as "*the deliberate consideration of all the circumstances concerned with a project in order to evolve the best method of achieving a stated objective*". This involves defining the project, setting the overall duration and identifying risks.

Planning is a vital element in any construction process without which it is impossible to envisage the successful conclusion of any project.

Planning considers:

- What to do.
- When to do it.
- How to do it.
- What resources are required to do it.
- Who is going to do it.
- How long it will take to do it.

The motto is –

PLAN YOUR WORK . . . AND WORK TO YOUR PLAN!

Planning may be considered on a long- or short-term basis, long-term planning including the development of a master plan for the whole project, whilst short-term planning is a much more detailed operation, which considers only the next three to four weeks of operations, or a particular element of the work. It is essential that each short-term plan should overlap the other by one to two weeks to allow for past and future activities to be considered.

The planning process considers each operation in turn, analysing the order and sequence of each operation, its dependency on previous operations, and the contractor's and client's resources to carry it out and the availability of those resources.

It is easier to track progress of operations using short-term, rather than long-term planning.

It is considered that there are three planning stages:

(A) PRE-TENDER PLANNING

This covers the planning considerations during the preparation of a cost estimate and its conversion into a submitted tender price.

The process enables the contractor to

- establish a realistic construction period on which the tender may be based. The client may set the starting date and the completion date for the work or may set the starting date and invite the tendering contractors to propose a completion date.
- consider the most effective economical construction methods, using the most effective resources.
- assist the build-up and pricing of contract establishment charges and equipment expenditure.

This is usually carried out in broad outline format as the contractor may not be successful in its tender.

Typical information which may be gathered and used at the tender planning stage includes:

1 *A brief description of the project and its location, details of the client, and the various consultants, key companies, third parties and individuals.*
2 *Details of the local planners and any planning issues or restrictions.*
3 *Access to site for delivery and collection of equipment, plant and materials and also removal of waste.*
4 *Details of existing services and the location of new and existing connection points. Details of the relevant service providers.*
5 *General details of the geography and topography of the area, past use, ground investigations, groundwater levels, and other local knowledge.*
6 *Location of tips and suppliers including quarries.*
7 *Availability of labour, equipment, plant and materials.*
8 *Details of possible Subcontractors.*
9 *Typical weather conditions.*

This information is then collected together with information given to the contractor at tender stage to develop a pre-tender plan. Time-based elements can then be transferred to a pre-tender programme, usually in a bar chart format which together with the method statements then aids the estimator in preparing the tender and assessing the need for establishment charges and major multi-use equipment such as craneage, scaffold and skips.

Clients should instruct contractors to submit programmes with their tenders, though it must be appreciated that as the contractor may not be successful with its tender it will serve as an indicative outline of how the contractor would carry out the works should it be successful.

The ECC provides, as an optional statement, for the contractor to submit a first programme for acceptance within a specified number of weeks of the Contract Date.

(B) PRE-CONTRACT PLANNING

This covers the planning considerations once the contract has been awarded. It is essential that this planning stage takes place prior to commencing work on the site.

The process enables the contractor to:

- establish a clear strategy for carrying out the works.
- comply with contract conditions and requirements.
- establish a construction sequence on which the master programme may be based.
- identify key dates by which parts of the work are required to be completed or the client is to provide information or resources to enable the contractor to complete work to the programme requirements.
- enable the assessment of contract budgets and cash flow forecasts.
- schedule key dates for procurement of key materials and subcontractor tendering, placing orders and commencement.
- schedule pre-contract meetings.

(C) PROJECT PLANNING

This is required to be implemented in order to maintain control and ensure that the project is completed on time and within the cost limits established at the tender stage.

The process enables the contractor to:

- monitor and update the master programme on a regular basis.
- optimise and continually review resources.
- keep progress under review and report on variances.

(ii) What is programming?

Programming is drawing all the planning considerations together into a present-able graphical format such as a bar chart and supporting tables which show the various operations, their durations, relationship between the operations, critical path and float, and the required resources so that it can be used as a project reporting and control mechanism, and form the basis of progress reporting to the client.

Programming is an essential part of any construction contract, but whilst other contracts often pay only a passing interest to the programme, the NEC3 family sees it as a vital tool from which to:

(i) *At tender stage, assess the contractor's ability to carry out the project.*
(ii) *Measure and report the contractor's actual progress against that planned.*
(iii) *Assess the effect of compensation events upon Completion and Price.*
(iv) *Ensure the Project Manager takes appropriate action when required.*

Preparation of a programme

The process of writing a programme normally consists of three stages:

(i) *Identifying the various operations to be carried out – the operations may initially be written as a list, identifying each operation in chronological order. That order may depend on natural construction sequence, constraints to access and resource availability.*

(ii) *Considering the duration of each operation – the duration of each operation can only be a forecast rather than an exact science as various issues have to be considered such as length of working days, differing productivity outputs, changes, resource availability, working conditions, changing weather and equipment breakdowns which can affect forecast durations. The contractor will then make appropriate adjustments as the work progresses.*

(iii) *The relationship between each operation will involve inserting links between operations which have a dependency on each other, e.g. "start to start", "finish to start", "finish to finish". From this can be established what the logical path is in order to execute the various operations, what operations need to start at the same time, what operations have to be completed before another operation can start, and what operations must finish at the same time.*

Figure 3.1 shows a simple programme for the construction of two storage tanks. One can interpret from the programme that there is an initial period for preparing the site, the construction of the foundations to both tanks, then commencing once the site has been prepared.

The construction of the tank foundations, the installation of the tanks themselves, and the installation of the pumps and tank painting for both tanks then progresses concurrently. Whilst the tanks are being painted, the external paving around the tanks can then progress, then the final operation – final testing – will progress with completion of this operation coinciding with completion of the external paving, and thus completion of the works.

The problem is that without any links, the dependency of the operations upon each other is not known, therefore if, for example, the construction of the foundations to Tank No. 1 is delayed, either by the contractor or the client, the effect on other operations will not be known.

Figure 3.2 includes links which show the logic and dependency of each operation, therefore, if the construction of the foundations to Tank No. 1 is delayed, it is on the critical path and therefore the delay to the foundations will have a direct delay effect on all the following operations.

The term programme is not defined in law or within contracts, the dictionary definition normally stating that a programme is "a plan or schedule of activities to be carried out". The ECC does not prescribe the form the programme submitted for acceptance must take (unless stated within the Scope); it would normally be a bar chart programme.

Before considering in detail the requirements of the contract, it is worth reviewing various forms of programme in common use.

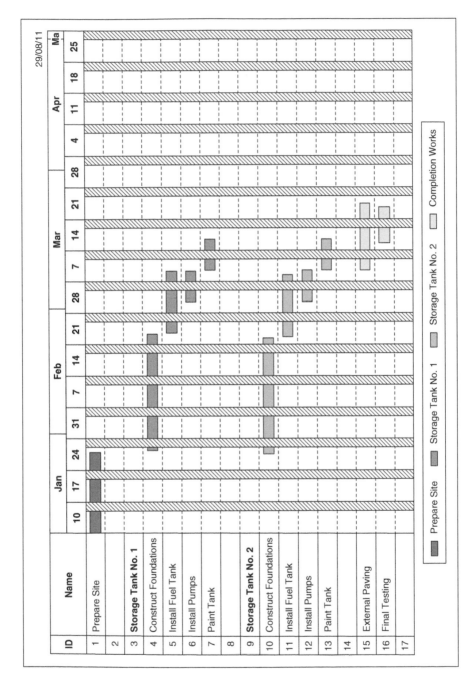

Figure 3.1 Programme with unlinked operations

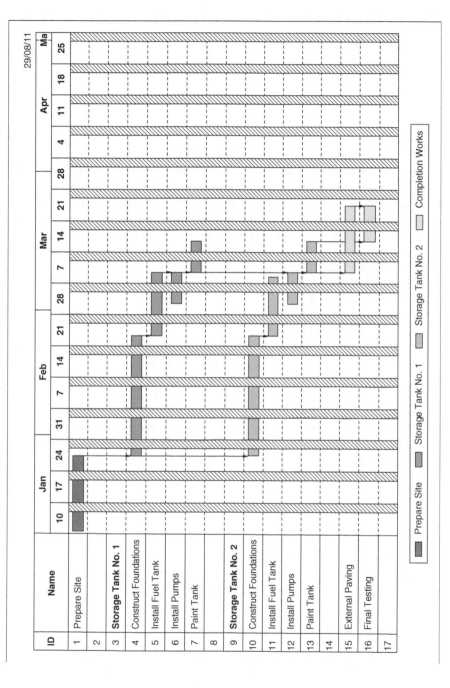

Figure 3.2 Programme with linked operations

1. Bar charts

Any programme, at any stage of a tender or project, is normally written as a bar chart or Gantt chart (Henry Laurence Gantt (1861–1919) US Engineer) showing operations which may occur either on or off site including their sequence and duration, with the start and finish of the bar depicting the start and finish date of the operation. Linked bar charts are the usual form of programme in common use on construction contracts. They are often supplemented by other information.

ADVANTAGES OF BAR CHARTS

- The simple format allows them to be readily constructed, they are user friendly and understood by most practitioners.
- They illustrate the specific activities, their durations and during the progress of the work, actual versus planned progress can be clearly shown.
- They can be used at all stages of the planning process through pre-tender, pre-contract and contract planning. They assist in showing the relationship between the pre-tender programme, the master programme and short-term programmes.
- They clearly relate to the construction sequence – the use of linking on bar charts aids the overlapping of related activities and can enable the critical path to be identified.
- They are easily updated, at weekly and monthly time intervals, for review purposes and progress reports.
- Key milestone symbols may be introduced to highlight critical dates with regard to key contract stages and information requirements.
- Resources may be readily related to the rate of working – labour histograms, value forecasts (value/time), and cumulative labour and plant forecasts in the form of project budgets.
- The programme forms the basis of financial forecasting for both the client and the contractor as a cash flow forecast can be prepared by pricing the programme.

DISADVANTAGES OF BAR CHARTS

- Without the use of links, they cannot properly depict the relationship and dependency between activities; for example, where an activity must be wholly or partially complete before the next can start.
- Complex interrelationships cannot always be clearly shown.
- They may give too simple a picture, which may be misleading.

2. Elemental trend analysis (line of balance) charts

The elemental trend analysis (line of balance) as with many programming methods originated from the manufacturing industry and is used where there is repetition of work either in number of units such as housing, or in repetitive floors in constructing or refurbishing multi-storey buildings. The unit or floor numbers are shown on the vertical y-axis, and the timeline of the project on the horizontal x-axis.

ADVANTAGES OF ELEMENTAL TREND ANALYSIS CHARTS

- They are very suitable for repetitive work and strict trade sequencing as the flow of work can be clearly demonstrated.
- They are more sophisticated than bar charts and simpler than network diagrams.
- They clearly show the rate of progress of each trade.

DISADVANTAGES OF ELEMENTAL TREND ANALYSIS CHARTS

They are totally unsuitable for non-repetitive projects and largely unsuitable for limited repetition.

- They are not easily understood, user friendly, or readily understood by all practitioners.

3. Project network techniques (precedence diagrams, network analysis, critical path analysis)

Project network techniques comprise a number of methods but all are related to the dependency and precedence of one activity over another.

Networks may be represented as arrow diagrams where activities are represented by arrows and events (start and finishes) by "nodes". Activities take time, and are represented as months, weeks, days or even hours, but events do not.

As stated earlier, the bar chart is a clear and simple method of plotting activities, but it has a clear disadvantage in that the relationship between activities cannot be properly represented. With networks, the relationship is immediately apparent and when a delay occurs specifically affected activities will stand out.

A project network illustrates the relationships between activities (or tasks) in the project, showing the activities either on arrows or nodes.

PREPARING NETWORK DIAGRAMS

The first operation as with any method is to compile a list of activities to be carried out.

One must then consider against each activity:

1   *Which activities must be completed before this activity can start?*
2   *Which activities cannot start till this one is completed?*

N.B. One should always consider when an activity can be done rather than when one thinks it will be done.

ADVANTAGES OF NETWORK DIAGRAMS

1   *Extremely effective when one is considering highly critical activities with strict dependency of several activities upon each other.*
2   *Float and critical path can be calculated fairly easily from such diagrams.*

DISADVANTAGES OF NETWORK DIAGRAMS

1   *Require a great deal of time and effort to produce them.*
2   *They are not practical for "normal" construction activities, which do not require highly detailed planning and monitoring of progress.*
3   *Do not provide a ready format for financial reporting.*

Most major projects are managed using software which combines high quality presentation with ease of revision and updating. This software can also link the master programme to short-term programmes and also resource schedules and cash-flow forecasts, however one must never forget the power of pen and paper.

3.3 The traditional approach of construction contracts

The value of a fully comprehensive construction programme is often misunderstood and undervalued by construction practitioners. This is often exacerbated by the minimal express requirement for a programme in most standard forms of construction contract.

Various commonly used forms merely state that the Contractor "submits its master programme for the execution of the Works", "states the order in which it proposes to carry out the Works" or "shall submit a detailed time programme . . . and shall submit a revised programme whenever the previous programme is inconsistent with actual progress or with the Contractor's obligations", often without going into any detail as to the required content and structure of the programme, the timing of the submission, the timing of the submission of any revised programme or any formal acceptance of the Contractor's programme.

The danger is that by understating the requirement for a programme within the contract, the true value of a programme is then misunderstood by the parties to that contract, and the programme submission and any acceptance can then descend into a "box ticking" exercise.

Clearly, although the programme requirement within these contracts is not spelt out in detail, it is advisable on any project for the client to require the Contractor to submit a programme, even though there may be no contractual requirement to do so, and it is also advisable for the Contractor to submit a detailed programme even if there is no contractual requirement to do so. The submitted programme should also identify the requirements to be met by the client.

The programme must be a dynamic tool which addresses all aspects of the project including design, information and other requirements, and the procurement process and lead-in times for sub-contractors and suppliers in addition to the construction operations. It must be flexible, and reviewed and if necessary revised on a regular basis as the work progresses and various events and changes come to light including the adjustments to float and time risk allowances.

3.4 The ECC approach

The ECC takes a different approach to other contracts in dealing with programme by clearly stating in detail what the Contractor is required to include in the programme it submits, with sanctions if it fails to do so, and also requiring that the Project Manager either accepts the programme, or gives its reasons for not accepting it.

The clear programme requirement enables it to be used to assess:

(i) *The Contractor's ability to carry out the project in terms of the intended resources and within the contractual time frame.*
(ii) *To demonstrate that all of the contract needs and requirements have been considered.*
(iii) *The Contractor's actual progress against that planned.*
(iv) *The effect of compensation events upon Completion and Price.*

3.5 Dates in the contract

There are a number of dates in the ECC which require some definition:

(i) Contract Date

The Contract Date is the date when the contract came into existence, however this is achieved, for example, by the client issuing a letter of acceptance to the Contractor. Note that there is no pro forma contract agreement within the contract, the principle being that parties will enter into contracts in various ways, using their own pro formas, standard letters or exchange of correspondence.

(ii) Starting date

In contrast with most forms of contract, the period of time for completion of the works is not stated. Instead, the starting date (and Completion Date) are decided and defined by the Client in the Contract Data. This has the advantage that tenderers know exactly when the works are required to start, and to complete.

The Contractor can start work from the starting date, but does not have access to the Site and cannot physically start on Site before the first access date (see below). It also carries risks from the starting date and so has to provide the insurances in the contract prior to that date.

(iii) Access date

The date(s), on which the Contractor may have access to and use of various parts of the Site, are stated in the Contract Data. Whilst the Contractor can start work on the contract from the starting date, it cannot start work on the site until the first access date.

There must always be at least one access date in the contract; if the Client has stated different access dates for different parts of the Site, then the Contractor must programme its activities according to the dates when it will be given access.

The Contractor may not require access on the dates stated in the Contract Data, in which case it should show on its programme the dates when it requires access. The Client is then required to give access to parts of the Site by the later of the access date in Contract Data Part 1, and the date for access shown on the Accepted Programme (Clause 33.1).

(iv) Completion Date

The Contractor is obliged to complete the works on or before the Completion Date as stated in the Contract Data or as may be revised in accordance with the contract. On or before the Completion Date, means that the Contractor can complete earlier than that date, but dependent on the statement within the Contract Data, the Client may not be willing to take over the works before the Completion Date.

The Project Manager decides the date of Completion, and is responsible for certifying Completion, as defined in Clause 11.2(2), within one week of that date.

Note that the Project Manager is not deciding when the Contractor has to complete by, as that is stated in the Contract Data and subject to any changes under the contract, but when it has completed.

Normally, the Contractor will request the certificate as soon as it considers it is entitled to it, but such a request is not essential.

Failure to complete on time constitutes a breach of the Contractor's obligation. If the Secondary Option X7 (Delay damages) is included, the Client is then entitled to pre-defined damages for that delay. If X7 is not included then unliquidated damages will apply.

Whilst many contracts use the terms "Practical Completion" and "Substantial Completion", the ECC does not, as these terms are often subject to various interpretations.

The ECC is more objective in defining Completion under Clause 11.2(2) as when the Contractor has

- *done all the work which the Scope states is to be done by the Completion Date and*
    This first bullet requires the Contractor to comply with the requirements of the Scope and supersedes the traditional terms "Practical Completion" or "Substantial Completion". This requirement may include, not only physical completion of the Works, but also submission of "as built" drawings, maintenance manuals, training requirements, successful testing requirements, etc. By including these requirements within the Scope, Completion has not been achieved until they are completed.

Contract: .... Contract No:	COMPLETION CERTIFICATE C/C No..

Completion achieved on: .... (Date)

The Completion Date is: .... (Date)

The *defects date* is: .... (Date)

The *defects on the attached schedule are to be*
 corrected within the defect correction period
 which ends on: .... (Date)

The *outstanding works on the attached schedule are to be*
 completed by: .... (Date)

Comments:

Signed: ....(Project Manager)	Date: ....

Copied to:

Contractor ☐   Project Manager ☐   Supervisor ☐   File ☐   Other ☐

Figure 3.3 Sample completion certificate

- *corrected notified Defects which would have prevented the Client from using the works or Others from doing their work.*

  The works may contain defects, though these are not significant enough to prevent the Client from practically and safely using the works. Completion is certified with the requirement that the Contractor corrects the defect before the end of the defect correction period following Completion.

There is also a fall back within Clause 11.2(2) in that if the work which the Contractor is to do by the Completion Date is not stated in the Scope, Completion is when the Contractor has done all the work necessary for the Client to use the works and for Others to do their work. This broadly aligns with the traditional terms "Practical Completion" and "Substantial Completion", where the Client is able to take beneficial occupation of the works and use them as intended.

(v) Key Date

A Key Date is defined in Clause 11.2(11) as the date by which a stated Condition has to be met. It is an optional clause allowing the Client, should it wish, to include dates when the Contractor must complete certain items of work, possibly to allow others to carry out other work. If used, the Condition and the applicable Key Date are defined by the Client in Contract Data Part 1.

"Key Dates" must be differentiated from "Completion" or "Sectional Completion", in that with Key Dates the Client does not take over the Works, whereas with Completion and Sectional Completion, it does. In this respect there are no pre-determined delay damages applied to the Contractor who does not meet a Condition by a Key Date.

However, if the Project Manager decides that the Contractor has not met the Condition stated by the Key Date, and the Client incurs additional cost on the same project as a result of that failure, then the Contractor will be liable to pay that amount (Clause 25.3).

By referring to "the same project", the Client cannot claim costs incurred on another project as a result of the failure of the Contractor to meet a Key Date. This cost is assessed by the Project Manager within four weeks of when the Contractor actually does meet the Condition for the Key Date.

Example

The Contractor is building a new outpatients department for a local hospital. The Client wishes to have new X-ray machinery installed by a specialist which it will employ directly, and who will install the machinery as the building work progresses.

In order to do that, the Contractor has to complete the part of the building that will house the machinery and it has to install the necessary electrical power facilities, so that the machinery can be tested prior to completion of the project.

The necessary work to be carried out by the Contractor in readiness for the specialist to carry out its part in order to meet the Condition will be stated, and this will have a Key Date attached to it.

N.B. The work should also be included within the Scope, the Key Date provision only stating what part of the work is required to meet the Condition stated.

If the Contractor then fails to meet the Condition by the Key Date and the Client incurs a cost in having to postpone the installation of the X-ray machinery, then this cost is paid by the Contractor. The cost must be incurred on the same project, so for example if the late installation of the X-ray machinery incurs a cost on another project, then this is not paid by the Contractor.

Note that this amount is the only right the Client has in respect of the Contractor's failure.

3.6 Programme requirements within the ECC

The ECC has quite comprehensive and specific requirements for a programme, in the form stated in the Scope, to be submitted by the Contractor to the Project Manager, these requirements being defined by nine bullet points within Clause 31.2.

It is not uncommon for a Contractor to hand the programme to the Project Manager at a progress or early warning meeting, but it is important that the submission be made in a formal manner, ideally using a standard template (see Fig. 3.4) recording what was submitted, the date of the submission, and the response.

Some practitioners say that the requirement within Clause 31.2 is very onerous and in some cases excessive, however it is critical that the programme shows in a clear and transparent fashion what the Contractor is planning to do, when and how long each operation will last, and what it needs for the Client and others to be able to comply with it.

There is nothing within the requirements of Clause 31.2 that a competent Contractor would not ordinarily include within a professionally constructed programme submitted for the benefit of itself and the receiving party. Whether and how it chose to show the information could be another matter!

There are two alternatives for the Contractor's submission of its first programme for acceptance:

(i) *It may submit, or be required to submit, a programme with its tender in which case it is referenced by the Contractor in Contract Data Part 2. Note that under Clause 11.1(1) "the Accepted Programme is the programme identified in the Contract Data or is the latest programme accepted by the Project Manager". Therefore, any programme referred to in Contract Data Part 2 and submitted by the Contractor as part of its tender, automatically becomes the first Accepted Programme, even though it is unlikely that it will comply with the requirements of Clause 31.2.*

Contract: .... Contract No:	CONTRACTOR SUBMISSION C/S No..

Section A: Submission of:

To: Project Manager/Supervisor   ☐ Drawings

                                            ☐ Programme

                                            ☐ Test Results

                                            ☐ Other

The following information is transmitted for your acceptance:-

Description   Date   No./Rev.   Copies

Response reqd. by ....

Signed: (Contractor) ....   Date ....

Section B: Reply

To: Contractor

The submission is returned:-

☐ Accepted   ☐ Accepted as noted

☐ To revise and resubmit   ☐ Rejected as noted

Notes:

Signed: (Project Manager)	Date:................

Copied to:

Contractor ☐   Project Manager ☐   Supervisor ☐   File ☐   Other ☐

Figure 3.4 Sample Contractor's submission

(ii) *If a programme is not identified in Contract Data Part 2, the Contractor submits a first programme to the Project Manager for acceptance within the period of time after the Contract Date, this period of time being stated in Contract Data Part 1. This will be a programme intended to comply with Clause 31.2, and is the first "meaningful" programme, even if one is submitted with the Contractor's tender.*

Such is the importance of a programme within the ECC that if no programme is identified in Contract Data Part 2, one quarter of the Price for Work Done to Date is retained (Clause 50.5) until the Contractor has submitted the first programme showing the information the contract requires.

Note that a Contractor who submits a first programme which shows the required information, but the Project Manager does not accept it, would not be liable for withholding of payment under this clause. The clause relates to the Contractor's failure to submit a programme showing the information the contract requires, not whether the Project Manager accepts it.

Clause 31.2

The *Contractor* shows on each programme which it submits for acceptance

- *the starting date, access dates, Key Dates and Completion Date*

  As stated previously, these dates are stated by the Client in Contract Data Part 1. With regard to Completion Date, if the Client has decided the completion date for the whole of the works then it is inserted into Contract Data Part 1. Alternatively, if the Contractor is to decide the completion date it inserts this into Contract Data Part 2 as part of its tender.
- *planned Completion*

  This is the date the Contractor plans to complete the Works; this is different to the date by which the Contractor is required to complete the Works as stated in Contract Data Part 1, or as may be revised in accordance with the contract.

  Although the Contractor can complete any time between its planned Completion date and the Completion Date, failure to complete on or before the Completion Date constitutes a breach of the Contractor's obligation and if Option X7 has been selected, the Contractor is liable to pay the Client delay damages.

  The difference between the dates when the Contractor plans to complete and when it is required to complete is float within the programme. This float, often referred to as "terminal" float, belongs to the Contractor and is attached to the whole programme. If a compensation event occurs, a delay to the Completion Date is assessed as the length of time that planned Completion is later than planned Completion as shown on the Accepted Programme (Clause 63.5).

  It is vital that, as planned Completion can change on a regular basis, the Contractor shows a realistic planned Completion in each revised programme submitted for acceptance, but also that the Contractor and Project Manager comply with the timescales in the contract for assessing and implementing compensation events, as it is not unusual for planned Completion to be shown later than the Completion Date leading one to believe the Contractor is in delay, yet the reason for this is either that the timescale in the contract is being adhered to but is giving a false picture as a programme is submitted whilst the Contractor is still preparing its quotation for a compensation event, or that compensation events have not been assessed and implemented in accordance with the timescales stated in the contract.

- *the order and timing of the operations which the Contractor plans to do in order to Provide the Works*

  This shows the sequence, duration and dates of the various operations the Contractor plans to do, and should include dependencies and links between the various operations. It should also show sequence, duration and dates in the procurement process, finalisation of information, placing orders, etc. It should also show work to be carried out off site.

- *the order and timing of the work of the Client and Others as last agreed with them by the Contractor or, if not so agreed, as stated in the Scope*

  The Contractor will state what it requires the Client and Others to do in order that the Accepted Programme can be met. These dates may already have been agreed with the Contractor, but may also have been inserted into the Scope by the Client. "Others" are parties outside the contract, i.e. not the Client, Project Manager, Supervisor, Adjudicator, the Contractor or a Subcontractor or Supplier to the Contractor. Examples of "Others" would be planning authorities, utilities companies, etc.

  This requirement is often provided in the form of an Information Required spreadsheet identifying what information is required and the latest date for providing it, failing which a compensation event can occur (Clause 60.1(5)). It is important to recognise that the programme is not just the bar chart but may include spreadsheets, schedules and graphs.

- *the dates when the Contractor plans to meet each Condition stated for the Key Dates and to complete other work needed to allow the Client and Others to do their work*

  If there are Key Dates in the contract (Clause 11.2(11)), then the Contractor must show how and when it intends to meet each Condition by these Key Dates. There may be provision for other Contractors directly employed by the Client to carry out work, for example a fit-out Contractor.

Provisions for

- *Float*

  Float is any spare time within the Contractor's programme, after time risk allowances have been included, and represents the amount of time that operations may be delayed without delaying the following operations (free float) and/or planned Completion (total float). It can also represent the time between when the Contractor plans to complete and when the contract requires him to complete (terminal float).

Float absorbs to a certain extent the Contractor's own delays or the delays caused by a compensation event, thereby lessening or avoiding any delay to planned Completion. In effect, no delay arises unless float on the relevant and critical operations reduces to below zero.

Programming is never an exact science, so float gives some flexibility to the Contractor in respect of incorrect forecasts or its own inefficiencies. As the work progresses the float will change as output rates change, Contractor's risk events take place and also compensation events arise.

The general belief, certainly amongst many Contractors, is that float belongs to them, as they wrote the programme, and therefore have the right to work to that programme and use any float that it contains. Conversely, Clients believe that if it is free time then they have the right to use it, but it actually depends where the float is and what it is for.

In principle float other than terminal float or time risk allowances is a shared resource, it is spare time, it belongs to the project, and therefore may be used by whichever party needs it first.

There are three types of float to consider:

(i) *the amount of time that an operation can be delayed before it delays the earliest start of the following operation (free float)*
(ii) *the amount of time that an operation can be delayed before it delays the earliest completion of the works (total float)*
(iii) *the amount of time between planned Completion and the Completion Date (terminal float).*

Generally, with other contracts, if the Contractor shows that it plans to complete a project early, then it is prevented from completing as early as it planned, but it still completes before the contract completion date, then although there may be an entitlement to an extension of time under the contract, for example exceptionally adverse weather, none will be awarded as there is no delay to contract completion, but it may have a right to financial recovery subject to its proving loss and/or expense.

The ECC deals with this in a different way. If the Contractor shows on its programme planned Completion earlier than the Completion Date and it is prevented from completing by the planned Completion Date by a compensation event, then, when assessing the compensation event, Clause 63.5 states "a delay to the Completion Date is assessed as the length of time that, due to the compensation event, planned Completion is later than planned Completion as shown on the Accepted Programme current at the dividing date", therefore any terminal float is retained by the Contractor, the period of delay being added to the Completion Date to determine the change to the Completion Date.

Any delay to planned Completion due to a compensation event therefore results in the same delay to the Completion Date. Therefore, the extension is granted on the basis of time the Contractor is delayed, i.e. entitlement, not on how long it needs to achieve the current Completion Date.

Similarly, with Key Dates, "a delay to a Key Date is assessed as the length of time that, due to the compensation event, the planned date when the Condition stated for a Key Date will be met is later than the date shown on the Accepted Programme current at the dividing date".

In assessing any delays to Completion Date and/or Key Dates only those operations which the Contractor has not completed and which are affected by the compensation event are changed.

Example

The Completion Date is Friday 16 February 2018, but the Contractor is planning to complete two weeks earlier, i.e. Friday 2 February 2018, the programme showing the last operation (roof covering, which has a duration of five days), being completed on that date. The Project Manager has accepted the Contractor's programme.

At the beginning of January 2018, the Project Manager gives an instruction to change the specification for the roof from concrete tiles to slates. This is a change to the Scope and therefore a compensation event under Clause 60.1(1).

The Contractor prepares its quotation and finds, when speaking to its roof slate supplier, that it cannot get the slates delivered to Site until week commencing Monday 5 February 2018, therefore the roof cannot be completed until Friday 9 February 2018; the Contractor then shows in its quotation a delayed to planned Completion of 1 week.

As Clause 63.5 states, "a delay to the Completion Date is assessed as the length of time that, due to a compensation event, planned Completion is later than planned Completion as shown on the Accepted Programme", assuming the Project Manager accepts the quotation, planned Completion becomes Friday 9 February 2018 and the Completion Date becomes Friday 23 February 2018.

So the Contractor is still completing two weeks earlier than the Completion Date.

- *Time risk allowances*

  Time risk allowances are essentially a form of float and are included within the duration of a specific operation. However, time risk allowances are normally "within an operation" rather than "between operations".

Essentially, a time risk allowance is the difference between what is the most "optimistic" (shortest) duration to complete an operation and what is a "realistic" (expected) duration, bearing in mind the risks that the Contractor may face in completing that operation. It is important that the Contractor's staff are fully informed as to how the time risk allowances are being presented, so there is no misunderstanding of how the works have been programmed and priced.

The Contractor's time risk allowances are to be shown on its programme but there are no requirements in the contract as to how these allowances should actually be shown, three possible methods being:

(i) *Show a single bar for an operation, within which the time risk allowance is shown. This is difficult to do with some planning software.*

(ii) *Show a separate bar which represents the time risk allowance for a specific operation. This however adds to the lines on the programme.*

(iii) *Show a separate column identifying the time risk allowance included within the total duration of the operation.*

Clause 63.8 states that the assessment of the effect of a compensation event includes risk allowances for cost and time for matters which have a significant chance of occurring and are at the Contractor's risk.

It follows that they should be retained in the assessment of any delay to planned Completion due to the effect of a compensation event. These allowances are owned by the Contractor as part of its realistic planning to cover its risks in pricing its tender and also in managing changes. They should be either clearly identified as such in the programme or included in the time periods allocated to specific activities.

Example

The Contractor has an operation in its programme for "piling".

It believes that if the ground conditions are reasonable, the weather is reasonable, and there are no mechanical problems with the piling rig, then the piling operation will take him eight weeks to complete.

However, the Contractor's knowledge and experience tells him that rarely is everything reasonable and almost inevitably it will encounter some delay, so a realistic timescale would be ten weeks.

It should then show a ten-week duration for the piling operation, within which two weeks are its time risk allowance.

It is important to recognise that time risk allowances belong to the Contractor and are retained in the assessment of any delay to planned Completion due to a compensation event.

- *Health and safety requirements*

   This will include time taken in complying with contractual and statutory requirements, but also in inducting new staff and operatives.
- *The procedures set out in the contract*

   This can include timescales for acceptance of design proposals, proposed subcontractors, etc.
- *The dates when, in order to Provide the Works in accordance with its programme, the Contractor will need*

   - *access to a part of the Site if later than its access date*
   - *acceptances*
   - *Plant and Materials and other things to be provided by the Client*
   - *information from Others.*

Generally, the Contractor should be given access to parts of the Site, acceptance of its design, subcontractors and programme in a timely manner to allow him to Provide the Works. However, in respect of access, the Contractor may be in delay and may therefore not require access on the date shown in the contract, there may also be certain operations for which there may be a long lead-in time, which the Client or Others may not appreciate and therefore the Contractor can include the requirement for acceptances and information at an early date.

This information may be highlighted as a milestone on the bar chart itself, or attached to it in the form of information required schedules, resource schedules, etc.

Anything submitted by the Contractor as part of its programme must clearly and simply show the relevant operations for which access, Plant and Materials or information are required, and the implications if they are not provided.

It is important that the Contractor and the Project Manager then manage the process by raising any concerns in respect of their provision as early warnings.

- *Regarding each operation, a statement of how the Contractor plans to do the work identifying the principal Equipment and other resources which will be used*

   The Contractor is required to provide a statement of how it intends to carry out each and every item on each programme it submits for acceptance.

   Whilst accepting that the contract requires this information, this can be a somewhat idealistic clause which requires the Contractor to comply with the daunting and in many cases somewhat unreasonable requirement to produce all the information in its first programme submitted for acceptance, and particularly on a large complex project with a long duration the Contractor may not yet know how it will carry out each operation and what resources it will use. It also may not have appointed a subcontractor for a part of the works so will not have method statements from the company who is actually carrying out the works.

It is also a significant task for the Project Manager to accept all the information the Contractor has submitted, so it may be appropriate in this respect for the Contractor and the Project Manager to agree a rolling programme of submission of method statements as the work progresses and the methods and resources may become clearer, perhaps each submission covering the next three or four months' operations.

Method statements have become increasingly important in the construction industry, particularly as health and safety legislation makes the parties more responsible for stating the methodology they are going to use in carrying out the works, and how they have provided for the various risks involved. For a programme to be meaningful it should always be linked to method statements so that the methodology of carrying out the works and also the resources the Contractor intends to use are clearly stated.

The contract does not specify the form these statements should take and in many cases a Contractor will spend a great deal of time and paper producing detailed statements, which can be almost meaningless. They may be presented as a written statement or in tabular form but they should clearly show how the Contractor intends to do the work, the main headings being:

  (i) *Title of each operation*
 (ii) *Description of method*
(iii) *Quantities (where relevant)*
 (iv) *Principal Equipment to be used with outputs and durations*
  (v) *Labour type and gang sizes with outputs and durations*
 (vi) *Plant to be installed*
(vii) *Materials to be used.*

N.B. Cost is not normally included in method statements.

When preparing method statements, the Contractor should always consider and show:

  (i) *What is to be done*
 (ii) *How it is going to be done*
(iii) *Who is going to do it*
 (iv) *Where it is to be done*
  (v) *When it is to be done*

This information will then initially allow the Project Manager to consider whether the Contractor is providing sufficient resources to Provide the Works within the time available, and also safely, but also as the programme is revised and methods of construction and resources used may change, the Project Manager is then kept up to date as to the Contractor's intentions.

This information will also prove invaluable when assessing compensation events.

- *Other information which the Scope requires the Contractor to show on a programme submitted for acceptance.*

    Again, this shows that the Scope is not just the drawings and specifications, but information on all that the Contractor is required to do in Providing the Works.

    From these requirements one can see that the programme under an ECC contract is not just the bar chart, but a collection of documents including method statements, risk assessments, resource analyses, etc. All of these are submitted by the Contractor as its programme, and all of these must be considered by the Project Manager in determining whether to accept the programme. Extreme care should be taken to only require documentation which has a purpose and use in the administration.

3.7 Activity Schedule based on Option A

Clause 55.3 of Option A requires that if the Contractor changes its planned method of working so that the activities on the Activity Schedule do not relate to the operations on the Accepted Programme, or corrects the Activity Schedule so that the activities on the Activity Schedule relate to the Scope, it submits a revision of the Activity Schedule to the Project Manager for acceptance.

Under Clause 55.4, a reason for not accepting a revision to the Activity Schedule is that

- it does not relate to the operations on the Accepted Programme
- any changed Prices are not reasonably distributed between the activities which are not completed or
- the total of the Prices is changed.

3.8 The Project Manager's response

Clause 31.3

Within two weeks of the *Contractor* submitting a programme to him for acceptance, the *Project Manager* either accepts the programme or notifies the *Contractor* of its reasons for not accepting it.

A reason for not accepting a programme is that:

- the Contractor's plans which it shows are not practicable;
    The Contractor could be planning to use a particular working method or piece of Equipment that the Project Manager believes would not be appropriate; or the Contractor has allowed a particular quantity of labour or a planned output, which the Project Manager believes is not practicable.
- it does not show the information which the contract requires;
    It may not show Key Dates or Access Dates, method statements, or any indication of float.

- it does not represent the Contractor's plans realistically;

  The Project Manager may not believe the Contractor can complete an operation within the time it has allowed, or it could be showing planned Completion, which the Project Manager believes is too early and therefore unrealistic.

- it does not comply with the Scope;

  The Contractor may not have allowed for a constraint within the Scope. Or the programme may simply be in the wrong format. For example, the Scope may prescribe that the programme has to be compiled and submitted using a particular brand of software, but the Contractor has used a different brand.

The contract requires that the Project Manager either accepts or does not accept the Contractor's programme and if it does not accept it, the Project Manager clearly states why it does not accept it.

There is no "implied acceptance" if the Project Manager fails to accept or not accept the Contractor's first programme or subsequent submitted programmes. If it does not accept the programme, and it is not for a reason stated in the contract that is in Clause 31.3, then it is a compensation event, though the Contractor is not prevented from commencing the works so it is difficult to envisage what the Contractor would be seeking to recover from the breach through the compensation event process.

It has been known for Project Managers to respond to the submission of the programme either by making no comment at all within the two weeks they have to respond, or by stating that the programme is accepted on condition that the Contractor makes certain amendments to it. Neither of these are either acceptances or non-acceptances.

Because of these issues of non-response, or at least vague responses, the NEC4 ECC now has a new provision within Clause 31.3 in that, if the Project Manager fails to reply to the Contractor's submission within the time allowed, the Contractor may notify to that effect. If the failure to respond continues for a further week after the Contractor's notification, it is treated as acceptance by the Project Manager. This will hopefully goad an "errant" Project Manager to do what the contract requires it to do!

Note that acceptance of a programme which shows the Contractor completing late whether deliberately or via the new provision, does not mean that the Project Manager accepts the delay, it merely acknowledges that the delay is reasonably reflected within the programme and if Option X7 is selected, the Contractor is still liable for delay damages.

The Accepted Programme:

- effectively provides an agreed record of the progress of the job and where the delays have come from;
- provides a realistic base for future planning by both the Contractor and Project Manager;

- is the base from which changes to the Completion Date are calculated;
- is the base from which additional costs are derived.

Because each operation has a method statement and resources attached to it, the change in resources can be calculated and hence the change in costs. Because information is so much more transparent, there is more scope for working together.

3.9 Revised programmes

The timing for submission of revised programmes for acceptance is defined by Clause 32.2:

- *within the period for reply after the Project Manager has instructed the Contractor to*
     For example, the Project Manager could instruct the Contractor to submit a programme to enable an issue to be discussed in an early warning meeting.
- *when the Contractor chooses to*
     Again the Contractor may feel it beneficial to submit a programme to be discussed in an early warning meeting.
- *at intervals no longer than that stated in Contract Data from the starting date until Completion of the whole of the works*
     This is the longest period for submission of a revised programme.

The revised programme shows:

- *the actual progress achieved on each operation and its effect upon timing of the remaining work*

The progress achieved may be shown on the programme itself, or appended to it in the form of a progress report. Operations which are not yet completed will normally be shown as percentage complete, being the progress achieved to date against the total work within the operation on the date the assessment is made. It is useful if the Contractor and the Project Manager agree the progress statement before the revised programme comes into being. The submitted revised programme is then an undisputed statement of fact.

The revised programme should reflect planned vs actual progress, including early or delayed start, early or delayed completion of each operation.

It is important that, if the Contractor states a percentage completed to date, it should show actual progress achieved rather than time expired to date. Also, the Contractor should not just make statements such as "operation on programme", as this may be misleading. An operation on programme does not necessarily mean that the operation is not delayed, for example, if the Contractor is expecting to complete an operation late, then stating the words "operation on programme" can be interpreted as either in accordance with the contract, or progressing late in accordance with its expectations.

It is also important to recognise that many operations may not progress on a "straight line" basis, for example, building a brick wall may initially involve work and resources in setting out and building corners, with the bulk areas following on.

- *how the Contractor plans to deal with any delays and to correct notified Defects*

  The Contractor should show in its revised programme any delays which may or may not be caused by him, and also time needed to correct its own Defects.
- *any other changes which the Contractor proposes to make to the Accepted Programme*

  An example of this could be that the Contractor re-sequences part of the work or proposes to use a different method or different Equipment to that specified by him in a previous method statement.

It is therefore very much a "living" programme. If the Contractor does not submit a programme which the contract requires, then the Project Manager can assess compensation events, without receiving a quotation from the Contractor (Clauses 64.1 and 64.2).

Note that the requirement within the NEC3 ECC to show in the revised programme "the effects of implemented compensation events" no longer exists in the NEC4 ECC.

In the opinion of the author, this is a welcome deletion as the previous requirement caused confusion, in that a compensation event is not "implemented" until the quotation has been accepted or assessed by the Project Manager.

Whilst the completion date is not changed until a compensation event has been implemented, there is a danger in only showing compensation events that have been implemented, so the revised programme was not reflecting reality.

3.10 Record keeping

One of the most important, but often understated elements of managing a construction contract and its progress and any difficulties, whether one is a Contractor or a Project Manager, is the keeping of concise and timely records. It is beneficial if the records are agreed by the Contractor and the Project Manager.

These are then undisputed records of fact. It is then only the interpretation of these facts which might subsequently be at issue.

A good record keeping system will enable the Contractor or the Project Manager to notify early warnings promptly, assist the preparation of programmes, enable payments to be made under Options C to E, and also prove invaluable in pricing and assessing compensation events, all of which have timescales applied to them, so readily available records are a must.

Records can include:

- Site diaries
- Site meeting minutes
- Photographs, including dates
- Labour returns
- Equipment returns
- Weather records
- Correspondence
- Financial accounts and records

Also, on the rare occasions when a dispute arises and is referred to adjudication or the tribunal, records are invaluable.

3.11 Instructions to stop or not to start work

Under Clause 34.1, the Project Manager has the authority to give instructions to the Contractor to stop, not to start work for any reason and may later instruct him to re-start it. The instruction would be a compensation event under Clause 60.1(4). However, if the instruction was required due to a fault of the Contractor, for example unsafe working, the Project Manager should instruct the Contractor that the Prices, the Completion Date and the Key Dates are not to be changed (Clause 61.4).

3.12 Take over

Once the Contractor has completed the works and the Project Manager has certified Completion, the Client takes over within two weeks. This occurs even if the Contractor completes the works early.

However, there is an optional statement within Contract Data Part 1 that "the Client is not willing to take over the works before the Completion Date". If this is selected, then if the Contractor completes the works early it still has responsibility for the care of the works until the Completion Date.

Whilst many contracts have express provision for partial possession by the Client, normally with the consent of the Contractor which shall not be unreasonably withheld, the ECC does not, though under Clause 35.2 the Client may use any part of the works before Completion has been certified.

The Client takes over the works when it begins to use it, except if the use is for a reason stated in the Scope, or to suit the Contractor's method of working.

Note that, if the Project Manager certifies take over of a part of the works before Completion and the Completion Date, then it is a compensation event under Clause 60.1(15) if the Client takes over the works before Completion but after the Completion Date then the Contractor is already late in completing and therefore that would not be a compensation event. There is no provision for the Contractor to give or refuse consent to the Client taking over the works.

The Project Manager certifies the date of take over, or partial take over within one week of it taking place.

3.13 Acceleration

Acceleration in many contracts normally means increasing resources, working faster, or working longer hours so that completion can be achieved by the Completion Date; in effect, the Contractor is catching up to recover a delay. However, within the ECC acceleration means increasing resources, working faster, or working longer hours so that Completion can be achieved before the Completion Date.

The ECC has always had provision for acceleration, but within the NEC4 ECC (Clause 36), there has been a redraft of the provision to give a more collaborative approach.

In the NEC3 ECC, the Project Manager may instruct the Contractor to submit a quotation for acceleration. It also stated any changes to any Key Dates to be included within the quotation.

In the NEC4 ECC, the Contractor and the Project Manager may propose an acceleration to achieve Completion before the Completion Date. If they mutually agree to consider the proposed change, the Project Manager instructs the Contractor to submit a quotation for acceleration. It also states any changes to any Key Dates to be included within the quotation.

In reality, the decision to consider acceleration is usually a collaborative decision between the Project Manager and the Contractor, before the Contractor is instructed to submit a quotation for acceleration, rather than the Project Manager unilaterally issuing an instruction to the Contractor to submit a quotation, so the new provision mirrors reality.

The Contractor provides a quotation within three weeks of being instructed to do so.

The quotation must include proposed changes to the Prices, and a programme showing the earlier Completion Date and Key Date. As with quotations for compensation events, the Contractor must submit details within its quotation, however the contract does not prescribe how the quotation is to be priced, e.g. based on the Schedule of Cost Components.

The Project Manager replies to the quotation within three weeks of receiving it. The reply may be either:

- a notification that the quotation is accepted, or
- a notification that the quotation is not accepted and that the Completion Dates and Key Dates are not changed.

Note that there is no remedy if the Contractor does not submit a quotation, or if the Contractor's quotation is not accepted by the Project Manager, in that it could make its own assessment, as it could with a compensation event. Acceleration can only be undertaken by agreement between the Project Manager and the Contractor and cannot be imposed on the Contractor or any assessment imposed upon it without its agreement.

When the Project Manager has accepted the quotation for acceleration, it changes the Prices, the Completion Date and any Key Dates and accepts the revised programme.

3.14 How should time be addressed for items which are detailed in the Scope but are not included in the Bill of Quantities?

The Contractor Provides the Works in accordance with the Scope (Clause 20.1), therefore it is assumed that it programmes the Works in accordance with the Scope. Clause 31.2 (Programme) refers to "Providing the Works" and "Scope".

Information in the Bill of Quantities is not Scope or Site Information (Clause 56.1). It does not tell the Contract what it has to do, but it must clearly be a fair representation of the scope of the Works to allow the Contractor to accurately price the Works. Although a Contractor will often use the quantities within the Bill of Quantities as a guide to calculate timescales, the bill is used as a basis for inviting and assessing tenders, so it is a pricing and payment document, i.e. it deals with money.

If an item is included within the Scope, but not included in the Bill of Quantities it is not in itself a compensation event, however, one must consider the following three clauses:

> 60.4   *A difference between the final total quantity of work done and the quantity stated for an item in the Bill of Quantities is a compensation event if*
>
> - *the difference does not result from a change to the Scope,*
> - *the difference causes the Defined Cost per unit of quantity to change, and*
> - *the rate in the Bill of Quantities for the item multiplied by the final total quantity of work done is more than 0.5% of the total of the Prices at the Contract Date.*
>
> *If the Defined Cost per unit of quantity is reduced, the affected rate is reduced.*

For example, a difference in quantity causes the Contractor to have to adopt an alternative method or sequence of working, provided also that the remeasured final value of the individual bill item is more than 0.5% of the total value of the Bill of Quantities at the time the contract came into existence.

> 60.5   *A difference between the final total quantity of work done and the quantity for an item stated in the Bill of Quantities which delays Completion or the meeting of the Condition stated for a Key Date is a compensation event.*

A change in quantity is not, in itself, a compensation event. The change has to delay Completion or the meeting of a Key Date.

60.6   The Project Manager corrects mistakes in the Bill of Quantities which are departures from the rules for item descriptions and for division of the work into items in the method measurement or are due to ambiguities or inconsistencies. Each such correction is a compensation event which may lead to reduced Prices.

This clause deals with *mistakes* in the Bill of Quantities, a change in quantity is not a mistake. An example of a mistake is that an item has been completely omitted from the Bill of Quantities, so that is a compensation event.

However, for the Completion Date to be changed, the Contractor would have to *prove* that Completion was delayed as a direct result of correcting that omission.

An example of Completion being delayed could be a *major* item missed from the Bill of Quantities which meant that when the Contractor used the Bill of Quantities as a guide to calculating timescales, the bill was fundamentally not a fair representation of the work to be carried out and therefore the omission "misled" the Contractor into calculating an incorrect time period for the Works. But it is not a straightforward entitlement and the burden of proof would have to rest with the Contractor.

A simple example is that the Scope (Drawings and Specification) showed that a house had to be constructed. However, the Bill of Quantities did not include for the tiling of the roof of that house.

Under Option B, the Contractor may well have a case for claiming that it had not priced the roof tiling, but I would not allow a claim that it had not programmed for it.

4 Quality management

4.1 Introduction

Quality Management is covered within the ECC by Clauses 40.1 to 46.2, and also by additional clauses within Main Options C, D and E.

Quality management is a major management function within the built environment industry. Unless a contractor can guarantee its client a quality product, whether it be a project or a service, it cannot compete effectively with others in the modern market.

Quality often stands alongside cost as a major factor in contractor selection by clients. To be competitive and to sustain good business prospects, quality systems which are efficient and evidential need to be designed and maintained.

The role of quality management for a contractor is not an isolated activity, but intertwined with all the operational and managerial processes of that company.

The modern concept of quality is considered to have evolved through three major stages over many years. These stages are as follows.

Quality control

The earliest and most basic form of quality management is quality control, the term being defined by an interpretation of its two elements – "quality" and "control".

Quality

The term "quality" is often used to describe prestige products such as expensive jewellery and motor cars. However, although applicable to these items, the term "quality" does not necessarily refer to prestigious products but merely to the fitness of the product or service to the customer's requirements. Quality is therefore described as meeting the requirements of the customer.

Control

The concept of being "in control", or having something "under control", is readily understood – we mean we know what we intend to happen, and we are confident that we can ensure that it does.

Quality control introduces inspection to stages in providing a product, ensuring that it is undertaken to specified requirements. The quality is measured by comparing the work actually carried out against the sample initially provided.

Inspection is the process of checking that what is produced is what is required. It is about the identification and early correction of defects.

The major objectives of quality control can be defined as follows:

(i) to ensure the completed work meets the specification
(ii) to reduce clients' complaints
(iii) to improve the reliability of services delivered
(iv) to increase clients' confidence
(v) to reduce delivery costs.

Quality control is primarily concerned with defect detection and correction. The main quality control technique is that of inspection and statistical quality control techniques (i.e. sampling) to ensure that the services delivered and the materials used are within the tolerances specified. Some of these limits are left to the inspector's judgement and this can be a source of difficulty.

The outcome of an inspection can be viewed in both objective and subjective ways:

(i) Objective
   That which is quantifiable and measurable – line, levels, verticality and dimensions.
   There are some precise quantified inspections including the testing of plant and machinery, pressure tests in pipework and tests on electrical installations.
(ii) Subjective
   That which is open to the inspector's interpretation, e.g. cleanliness, fit, tolerances and visual checks.

Quality control implemented in construction

Traditionally there are two sets of documents that are used to determine the required quality of a construction project. These are the Specifications and the Contract drawings. The contractor uses these two documents during the site operations stage of any project to facilitate "quality" construction.

The process of construction is dissimilar to that of a production line, in that there are no fixed physical and time boundaries to each operation of the process, hence the positioning and timing of quality inspection cannot be predetermined. In construction quality checks are undertaken as each operation or sub-operation is completed.

The majority of quality checks are undertaken visually. Visual quality checks of each section of construction are undertaken by the contractors' engineers and foremen and then by the resident engineers and inspectors to ensure it complies with the drawings and specification.

Quantifiable quality checks are also made during the construction stage. These include testing the strength of concrete cubes, checking alignment of brickwork, and commissioning of services installations. The results of these quality checks are recorded and passed to the appropriate authority. The weakness of quality control is the development of the inspection mentality or culture whereby the construction contractors' operatives and engineers set their standards to that which they can "get past the inspector". In addition to potentially surrendering the standards of workmanship to an inspector it exposes the contractor to expensive re-work if the standards of workmanship obtained do not meet with the inspector's approval.

It would be much better if the contractors' engineers and operatives had a clear understanding of the quality required, were able to recognise it themselves, and were able to achieve it first time or regulate it by self-inspection. This concept being the basis of quality assurance potentially reduces the risks of producing unsatisfactory work and being involved in expensive re-work.

Notwithstanding the existence of quality assurance and the emergence of total quality management most clients still engage inspectors through their resident engineers or architects to reassure themselves. However, the impact and importance of the clients' inspectors is much reduced in a quality assured or total quality managed company.

Quality assurance

Whereas quality control focuses on defect *detection* and *correction* during the works, quality assurance is based on defect *prevention*.

Quality assurance is the process of ensuring that standards are *consistently* met, thereby *preventing* defects from occurring in the first place.

"Fit for purpose" and "right first time, every time" are the principles of quality assurance and the reference point for quality assurance is the International Quality Standard ISO 9000 family of standards. To be certified as operating to the ISO 9000 standard is now virtually seen as essential in today's construction industry. Many clients simply will not do business with companies not certified to ISO 9000.

Quality assurance concentrates on the delivery methods and procedural approaches to ensure that quality is built into the system.

Quality assurance is described under the following headings:

- evolution of quality assurance from quality control
- definition of quality terms
- quality assurance standards
- developing and implementing a QA system
- quality assurance in construction.

Total quality management

This is the third stage, and is based on the philosophy of continuously improving goods or services.

The key factor is that everyone in the company should be involved and committed from the top to the bottom of the organisation.

The successful total quality managed company ensures that their goods and services can meet the following criteria:

(i) Be fit for purpose on a consistently reliable basis
(ii) Delight the customer with the service which accompanies the supply of goods
(iii) Supply a quality service that is so much better than the competition that customers want it regardless of price.

It may be argued that successful service companies have to meet at least two of these criteria to stay successful. The pursuit of total quality is seen as a never-ending journey of continuous improvement.

4.2 Quality management system

Clause 40.1 requires the Contractor to operate a quality management system as stated in the Scope, submitted to the Project Manager for acceptance, and complying with the requirements stated in the Scope. This provision is new to the NEC4 Engineering and Construction contract, but has previously featured, in particular, within the NEC3 Professional Services Contract (and now in the NEC4 Professional Service Contract).

A quality management system is a set of interrelated or interacting elements that companies and other organisations use to direct, manage and control how quality policies are implemented and quality objectives are achieved.

A process-based quality management system uses a process approach to manage and control how its quality policy is implemented and quality objectives are achieved.

The quality management system will normally be to ISO 9000 standards, ISO 9000 defining a Quality Management System as "co-ordinated activities to direct and control an organization with regard to the degree to which a set of inherent characteristics fulfils the requirements".

ISO 9001 sets out the basic requirements for a quality management system, generally that it enables a company to consistently provide products or services that enhance customer satisfaction whilst meeting applicable statutory and regulatory requirements. It is essentially about providing quality assurance to customers.

It can be used by any organisation, large or small, regardless of its field of activity. In fact ISO 9001:2015 is implemented by over one million companies and organisations in over 170 countries.

Quality management principles

The standard is based on a number of quality management principles, including a strong customer focus, the motivation and implication of top management, the process approach and continual improvement. Using ISO 9001:2015 helps ensure that customers get consistent, good quality products and services, which in turn brings many business benefits.

A quality management system is intended to help a company to:

- identify, develop and implement efficient management systems.
- reduce waste to acceptable levels.
- ensure effective team working.
- regularly audit and review the Quality Management system to identify excellence, problems and areas for improvement.

It promotes customer loyalty by:

- ensuring customers' needs are fully identified and understood.
- ensuring project requirements are identified, understood, documented, and agreed by all.
- providing a management system that ensures on time delivery of the agreed product or service.
- providing brief, user friendly and easily accessible methods to express satisfaction or dissatisfaction and ensure that any dissatisfaction is resolved.
- meeting the national statutory and regulatory requirements.

It provides a good working environment and culture for staff by:

- promoting a culture of honesty, timely communication and assistance to each other.
- providing brief, user friendly and easily accessible procedures and processes that reflect the users' preferred method of working wherever practicable.
- ensuring that recognition is given to deserving staff.

Audits

Checking that the system works is a vital part of ISO 9001:2015. An organisation must perform internal audits to check how its quality management system is working.

An organisation may decide to invite an independent certification body to verify that it is in conformity to the standard, but there is no requirement for this. Alternatively, it might invite its Clients to audit the quality system for themselves.

4.3 Quality policy statement

A Contractor's quality policy statement is required by Clause 40.2 to be submitted by the Contractor to the Project Manager for acceptance. The quality policy statement defines the Contractor's management's commitment to achieving and maintaining quality. It is often just a single paragraph or an executive summary, but it should clearly state the Contractor's general quality orientation and clarify its basic intentions.

Again, this is new to the NEC4 Engineering and Construction contract, but has previously featured in particular within the NEC3 Professional Services Contract.

Quality policy statements should be used to generate quality objectives and should serve as a general framework for action. Quality policies can be based on the ISO 9000 Quality Management Principles and should be consistent with the organisation's other policies.

A typical quality policy statement could say:

> *"ABC Contractors are committed to achieving and consistently exceeding its Client's expectations, getting it right first time every time, by providing contracting services to the highest quality and within a safe and environmentally friendly manner. Our success in achieving these objectives is continuously measured through our quality management system accredited under ISO 9001:2015."*

Quality plan

The Contractor is also required by Clause 40.2 to submit a quality plan to the Project Manager for acceptance. The quality plan is a document that is used to specify what procedures and processes will be followed, and who is responsible for managing them, to ensure that the quality expectations of a particular Client will be met.

Again, this is new to the NEC4 Engineering and Construction contract, but has previously featured in particular within the NEC3 Professional Services Contract.

The quality plan may consist of the following sections:

- The Client's quality expectations – what the contract requires.
- Acceptance criteria – how quality will be measured.
- Quality responsibilities – who is responsible.
- References to any quality standards – what other codes of practice, legislative and/or other standards apply.
- Quality controls and audit processes – how the procedures and processes will be checked for compliance.
- Change management and procedures – how changes will be incorporated into the plan.

The plan should then be signed by all relevant parties to confirm that they have read and fully understand their responsibilities.

The degree to which the Client checks conformity with the statement and plan is entirely a matter for the Client. However, under Clause 40.3, if the Project Manager discovers a non-compliance, the Contractor is required to correct such failure. Such instruction would not be a compensation event.

The contract includes the provisions for carrying out tests, inspections and correction of defects, and also the new provision for a quality management system, including a quality policy statement and a quality plan to be provided to the Project Manager.

4.4 Defining Quality within the Scope

Let us now consider what the ECC requires in terms of Quality Management, which is defined by the Scope.

The Contractor is required to Provide the Works in accordance with the Scope.

The Scope must therefore define clearly what is expected of the Contractor in Providing the Works. It must cover all aspects of the work, either specifically or by implication, otherwise it may be deemed to be excluded from the contract.

A comprehensive description should therefore be considered for the Scope, which should be supplemented by specific detailed requirements in the form of drawings and specification. If reliance is placed solely on the scope of works, there is a danger that an item may be missed and could be the subject of later contention.

The Scope is the document that a tenderer can look to, to gain a broad understanding of the scale and complexity of the job and be able to judge its capacity as a Contractor to undertake it. It is written specifically for each contract and should be comprehensive, but it should be made clear that it is not intended to include all the detail, which will also be contained in the drawings, specifications and schedules.

Note that there is no hierarchy or precedence between documents within the Scope and in fact within the contract. If there is any ambiguity or inconsistency in or between the documents which form the contract, the Project Manager or Contractor notifies the other, and the Project Manager then states how the inconsistency or ambiguity should be resolved (Clause 17.1).

If the Scope has to be changed then this is a compensation event (60.1(1)) which is assessed as if the Prices, Completion Date and Key Dates were for the interpretation most favourable to the party which did not provide the Scope.

The Scope will normally be comprised of the following:

(i) Description of the Works

The Scope must include a clear definition of what is expected of the Contractor in Providing the Works. It must cover all aspects of the project, either specifically or by implication, otherwise it may be deemed to be excluded from the contract.

A comprehensive description should therefore be considered, which should be supplemented by specific detailed requirements in the form of a specification. If reliance is placed solely on the description of the works, there is a danger that an item may be missed and could be the subject of later contention.

(ii) Drawings

The drawings should detail all the work to be carried out by the Contractor and be complementary to the description of the works, and the specification.

(iii) Specification

The Specification is a written technical description which describes the character and quality of services, materials and workmanship, for work to be executed. It may also lay down the sequence in which various portions of the project are to be executed.

As far as possible, it should describe the outcomes required, rather than how to achieve them, criteria or standards which are required for the work, and complementary to the other constituents of the Scope and any notes or other text on the drawings, and the provisions of the contract.

(iv) Responsibilities of the parties

The responsibilities of the contracting parties and their representatives should be clearly defined in a manner which will leave no doubt as to the obligations that each is accepting, whilst also considering the written terms of the conditions of contract. This will also allow the procedures necessary to enable the contract to progress satisfactorily from inception to completion to be established at the outset of the service.

The contract as a whole will also define the responsibilities for the planning of the works, timings, resources, the supply of any free issue materials, the issuing of instructions and the form which these instructions are to take, the programming of the services, the method of measuring and evaluating the services, the circumstances which will constitute a variation to the scope of works and the duties of the parties during the contract.

4.5 Testing and inspections

The Contractor and the Supervisor are required to inform the other of tests and inspections before they start, and also the results of those tests. If a test shows that any work has a Defect, the Defect is corrected and the test repeated.

Figure 4.1 shows a sample Supervisor/Contractor Notification which may be used for this purpose.

The Project Manager assesses the cost incurred by the Client in repeating a test or inspection after a Defect is found and the Contractor pays the amount assessed.

The Supervisor is also obliged to do its tests and inspections without causing unnecessary delay to the work, or to a payment which is conditional upon a successful test.

The term "unnecessary delay" is subjective; clearly it must be expected that there may be some delay while the Supervisor does its tests and inspections.

Contract:	SUPERVISOR/CONTRACTOR NOTIFICATION
Contract No:	S/CN No...

To: Contractor

Type of Notification/Instruction:	Date	Time
☐ Notification of test or inspection	....	....
☐ Notification of result of test or inspection	....	....
☐ Notification of identified Defect	....	....

Instruction to search (Supervisor only):
Location and Details:

Result of test or inspection:
Test: PASSED/FAILED
Reason for failure:

It is planned to correct the Defect and retest on (date) ....

Defect corrected Yes/No   Date:........................
Other comments or action:

Signed: (Supervisor/ Contractor)	Date: ....

Copied to:

Contractor ☐   Project Manager ☐   Supervisor ☐   File ☐   Other .... ☐

Figure 4.1 Sample Supervisor/Contractor Notification

If the test or inspection causes unnecessary delay, then it is a compensation event (Clause 60.1(11)).

Whereas most of the interactions within the contract are between the Contractor and the Project Manager, within Section 4 they are between the Contractor and the Supervisor, the Project Manager playing a fairly minor role, only being involved in respect of arranging access for the Contractor to correct a Defect, assessing costs incurred by the Client in repeating a test or inspection and assessing the cost to the Contractor of correcting a Defect when it is not given access after the Client has taken over the works.

The Supervisor, in conjunction with the Contractor, notifies tests as and when required and carries out the various inspections as required by the contract. It is important to recognise, as previously stated in Chapter 1, that the Supervisor may be appointed from the Client's own staff or may be an external consultant, but it represents the Client in respect of testing and defects, it is not an assistant or deputy to the Project Manager.

Under Options C, D and E, when the Project Manager assesses the cost incurred by the Client in repeating a test or inspection after a Defect is found, it does not include the Contractor's cost of carrying out the repeat test or inspection.

Note, however, that under Options C, D and E (see below), the Contractor may be paid for correcting the Defect.

The Supervisor also has the right to watch any test being done by the Contractor.

If a payment is conditional upon a test or inspection being successful it becomes due for payment at the later of the defects date and the end of the last defect correction period if:

- the Supervisor has not done the test or inspection, or
- the delay to the test or inspection is not the Contractor's fault.

Should the contract require Plant and Materials to be tested before delivery, for example structural or electrical testing, they are not brought to the Working Areas until the Supervisor has notified the Contractor that they have passed the test or inspection.

4.6 Defining testing within the Scope

Testing, within this contract is defined through the Scope and the applicable law.

Any materials facilities and samples for testing, including who provides them, are also stated within the Scope.

The main documents within the Scope in respect of testing and inspection will again normally be the drawings and specification.

Testing can be related to individual Plant and Materials within the Working Areas, factory testing outside the Working Areas, testing of individual sections of work, welds and joints once they have been assembled or a multitude of mechanical and electrical tests.

Whilst the Scope will set out the procedures and parameters for testing and commissioning, the contract will set out the Parties' contractual obligations, and the consequences of failure to meet them. Many activities, for example electrical testing, will have to comply with legislation independent of the contract.

Performance testing is usually applicable where the design of Plant and Materials is the responsibility of the Contractor. This can include the operation of the completed plant and the checking of product quality and quantity, fuel consumed, use of energy, waste and by-products, environmental conditions and other aspects such as may be required. In such instances time limits for the rectification of performance defects must be stated together with penalties for failure.

The documentation required in the form of test certificates, warranties and guarantees, to enable proper records to be maintained should be stipulated, as will the effect in relation to the guarantee or warranty period and the extent to which it will apply. Many such records are only maintained electronically so access to the relevant asset management systems should be specified for the Client together with the requirements for data transfer at expiry of the service period.

Liability for defects should be defined in the contract together with the period of time for which such liability is to apply and the length of time within which defects are to be remedied.

It is imperative that the Scope, produced at tender stage and identified in Contract Data Part 1, defines the full extent and timing of tests and inspections required and the subsequent correction of any defects.

Testing is very often not covered in sufficient detail within the Scope, so it is important for the compiler of the Scope to consider a number of key questions:

(i) What is the purpose of the test?

What is it intended to prove or disprove? Is the test required before a following activity can be carried out by the Contractor or another party, or is it required before the Client can use a building or part of the building?

(ii) What is to be tested?

What elements or components are required to be tested? Is it to be tested in an assembled or unassembled state?

(iii) How is the test to be carried out, and who is to provide the materials, facilities and equipment to do the test?

Whether the test is a well-known "industry standard" test or not, the exact nature of the test must be clearly defined. Who supplies the equipment, who pays for the power or fuel for the equipment, etc? It is also important to detail who is to provide the materials and facilities to do the test and who meets the costs.

(iv) Who should do the test? Who should be present and who should be invited should they wish to attend when the test is taking place?

There may be a requirement for an independent party to carry out the test and provide results.

(v) When and where will the test take place? On site or off site?

Clearly state at what stage or on what date the test must take place. Sometimes, tests may have to be carried out on a periodic basis, for example on concrete samples on delivery and as the concrete cures.

(vi) What is the expected result or outcome from the test?

What is the test expected to prove? What results are needed to be gathered once the test has been carried out?

(vii) Who should be advised of the outcome of the test?

The Project Manager will normally be advised of the outcome of a test, but sometimes a subcontractor, or a third party outside the contract may need to be informed.

(viii) What should be done in the event of a successful test, e.g. certification?

Does a certificate have to be issued following successful completion of a test? Does that successful outcome mean that the Contractor can proceed to the next stage of the service? Is payment to the Contractor dependent on successful completion of the test?

(ix) What should be done in the event of an unsuccessful test, e.g. rectification or replacement, and further testing.

Does the Contractor have to repeat the test? Is the re-test done immediately after the first test, or should there be a period before the next test? Is it possible that even though something has failed a test, it may still be considered acceptable, possibly by the Contractor offering a cost saving?

(x) What about additional testing, i.e. to confirm a suspected defect or the extent of a suspected defect?

There may be a requirement for Plant and Materials to be tested off site before delivery; this is covered by Clause 42.1.

4.7 Defining a Defect

A Defect is defined within the contract (Clause 11.2(6)) as:

• *a part of the works which is not in accordance with the Scope*

The Scope provides the reference point for what the Contractor has to do to Provide the Works. If the work done by the Contractor does not comply with the Scope, then unless the Defect is accepted (Clause 44), the Contractor is obliged to correct the Defect.

In most cases the work done will fall short of the Scope, but as the words "not in accordance with" are used, work which exceeds the Scope requirement would also be a Defect. Clearly, the Project Manager may wish, having discussed with Client, to accept such a Defect!

A Defect may not necessarily mean that the work is not fit for purpose, in that the Defect could simply be a colour variation, for example the Scope stated a certain shade of red paint and the Contractor used a different shade of red paint, in which case a Defects exists as that part of the works was not in accordance with the Scope. In such a case, unless the shade of red is a particular corporate colour or a stipulation of a regulatory authority it may be prudent to accept the Defect.

It is also possible that work is "defective" but as it complies with the Scope, it is not a Defect! For example, the specification within the Scope may stipulate use of a particular material which then fails. In that respect the work is defective but as it complies with the Scope, it is actually not a Defect.

- *a part of the works designed by the Contractor which is not in accordance with the applicable law or the Contractor's design which the Project Manager has accepted.*

If the Project Manager accepts the Contractor's design under Clause 21.2, then subsequently the Contractor either does not comply with the applicable law, or changes the design, then again that is a Defect. If the Contractor wishes to change its design then it has to re-submit the new design to the Project Manager for its acceptance.

4.8 Searching for Defects

Under Clause 43.1, the Supervisor (not the Project Manager) has the authority to instruct the Contractor to search and gives reasons for the search with the instruction.

This action is normally defined as "uncovering" or "opening up" in other contracts. It is usually required in order to investigate whether a defect exists, and possibly the cause, and the correcting measures required.

Searching can include uncovering, dismantling, reassembly, providing materials and samples and additional tests which the Scope did not originally require.

If the Supervisor instructs the Contractor to search for a Defect and no Defect is found this is a compensation event (Clause 60.1(10)). However, if the search is needed only because the Contractor gave insufficient notice of doing work obstructing a required test or inspection, then it is not a compensation event.

4.9 Notifying Defects

Until the defects date, the Supervisor is obliged to notify the Contractor of every Defect as soon as it finds it, and the Contractor is obliged to notify the Supervisor as soon as it finds it.

Figure 4.2 shows a sample Defects Notification which may be used for this purpose.

Whilst the requirement for the Supervisor to notify the Contractor of each Defect is useful as the Contractor may not have been aware of the Defect or that it was a Defect, so it can correct it immediately, the obligation for the Contractor to notify the Supervisor of each Defect as soon as it finds it is somewhat cumbersome as the Supervisor probably does not want or need to know about every Defect, particularly if the Contractor is already correcting it.

| Contract: .... | DEFECTS NOTIFICATION |
| Contract No: .... | D/N No... |

The Contractor/Supervisor is notified of the following defects:

<u>Description</u>   <u>Date Corrected</u>

<u>The defect correction period is:</u>

Certified all defects
corrected................................(Supervisor)   Date: ....

Copied to:

Contractor ☐   Project Manager ☐   Supervisor ☐   File ☐   Other .... ☐

Figure 4.2 Sample Defects Notification

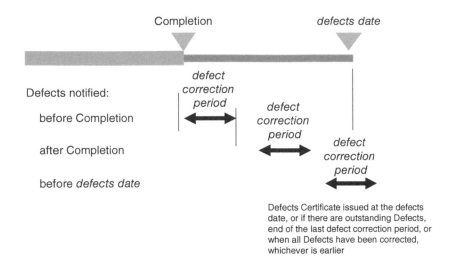

Figure 4.3 Defects Correction

4.10 Correcting Defects

Under Clause 44.1, the Contractor corrects a Defect whether or not the Supervisor notifies him of it.

The ECC has two periods for correction of Defects which are identified within Contract Data Part1.

Figure 4.3 is an illustration of periods for correcting Defects.

1   The period between Completion and the defects date for Defects which were not apparent or were not notified at Completion, or were notified, but did not prevent the Client from using the works. This period is set within Contract Data Part 1, and is normally 52 weeks. This is similar to the "defects liability period", "rectification period" or "defects notification period" in other contracts.

2   The *"defect correction period"* is the period within which the Contractor has to correct a notified Defect, failing which the Project Manager assesses the cost incurred to the Client of having the Defects corrected by other people and the Contractor pays this amount.

Note that, in this respect, although the Contractor has already paid the amount, the Client may wish to leave the Defect uncorrected. The defect correction period starts at Completion for Defects notified before Completion and when the Defect is notified for other Defects. The defect correction period is not effective prior to Completion, therefore if a Defect becomes apparent prior to this it has an obligation to correct it in readiness for Completion.

Some Clients may wish to designate different defects correction periods for different defects depending on their urgency. Contract Data Part 1 provides for this – the typical example defect correction periods for a water company upgrading domestic water supplies being:

Type A Defects – 4 hours

- Major leaks
- Defects related to potential health and safety issues

Type B Defects – 24 hours

- Minor leaks

Type C Defects – 7 days

- Cosmetic defects
- Defects which are not Type A or B

Defects which may prevent the Client from using the works must be corrected before a Completion Certificate can be issued (Clause 11.2(2)).

If a Defect is required to be corrected after the Client has taken over the works, the Project Manager arranges for the Client to allow access to the Contractor. The defect correction period in respect of the Defect does not commence until the necessary access has been provided.

4.11 Limitation periods

Whilst the ECC has a period from Completion to the Defects Date during which the Contractor is liable for correcting Defects, most countries also have a statutory limitation period for latent defects which come to light at a later date, beyond which a claim cannot be made for defective work.

In the UK, the Limitation Act 1980 governs time limits for bringing different types of legal claims, for example, the time limit for actions founded on simple contract is 6 years and for a specialty contract or deed, the time limit is 12 years; in both cases this is measured from the date on which the cause of action accrued.

4.12 The Defects Certificate

The Defects Certificate is issued by the Supervisor, to the Project Manager and the Contractor at the defects date if there are no notified Defects, or otherwise at the earlier of:

- the end of the last defect correction period
- the date when all notified Defects have been corrected.

Some practitioners state that as the Project Manager issues the Completion Certificate, it should also issue the Defects Certificate, which is common with most other contracts, but by requiring the Supervisor to issue the Defects Certificate, the ECC reinforces the role and responsibility of the Supervisor in dealing with Defects.

Note the definition of a Defects Certificate under Clause 11.2(7):

> *The Defects Certificate is either a list of Defects that the Supervisor has notified before the defects date which the Contractor has not corrected or, if there are no such Defects, a statement that there are none.*

Whilst many contracts require any defects to be corrected before the Defects Certificate, or its equivalent is issued, under the ECC, the Defects Certificate is issued by the Supervisor at the appropriate time and may either list Defects which the Contractor has not corrected, or a statement that there are no outstanding defects.

Note that the Defects Certificate is not conclusive, in that if a Defects Certificate states that there are no defects it does not prevent the Client from exerting its rights should a defect arise later or the Supervisor did not find or notify it. However, the Client may wish to take action against a Supervisor who did not carry out the level of inspection it had paid him to do!

The Supervisor will normally only list what are usually defined as "patent defects", i.e. those which are observable from reasonable inspection at the time, examples being a defective concrete finish or an incorrect paint colour and may not include what are usually defined as "latent defects", which may be hidden from reasonable inspection and may come to light at a later date, examples being some structural defects.

The Contractor's and others' liability for correction of latent defects and other costs associated with them will be dependent on the applicable law, and liability may remain despite the issue of the Defects Certificate.

4.13 Accepting Defects

As stated above, a Defect is defined as "a part of the works that is not in accordance with the Scope or a part of the works designed by the Contractor which is not in accordance with the applicable law or the Contractor's design which the Project Manager has accepted (Clause 11.2(5)).

In the event that a Defect becomes apparent, there are two options:

(i) The Contractor corrects the defect (Clause 44.1) so that the part of the Works is in accordance with the Scope and the design is in accordance with the applicable law or the Contractor's design as accepted by the Project Manager.

(ii) The Contractor and the Project Manager may each propose to the other that the Scope should be changed so that a defect does not have to be corrected (Clause 45.1). This is normally applied to "cosmetic defects" rather than any with structural or other major effect.

Note that "each proposes to the other" requires the acceptance of the Contractor and the Project Manager, the latter probably discussing the matter with the Client. The Contractor submits a quotation to the Project Manager for reduced Prices or an earlier Completion Date, and if accepted, the Project Manager changes the Scope, the Prices, the Completion Date and any Key Date and accepts the revised programme.

Example

The Contractor has carried out a large area of wall tiling in a proposed new railway station.

Whilst visually the quality of the work appears to be very good, the tiles have not been laid to the required tolerance within the Scope and therefore the work is technically defective.

The railway station is due to be completed in two weeks, so if the Contractor has to take down the wall tiles, re-order another batch of tiles and then lay the new tiles to the required tolerance, that could take a considerable time and possibly prevent completion of the works, or at least that specific part of the works.

In this case, provided both the Contractor and the Project Manager are prepared to consider changing the Scope, then the Contractor provides a quotation to the Project Manager stating a financial saving (reduced Prices), based on him not having to correct the Defect or an earlier Completion Date or both, and if the Project Manager accepts the quotation the Project Manager changes the Scope, the Prices, the Completion Date and any Key Date and accepts the revised programme.

This is obviously not an option where the design or a part of the works is not in accordance with the applicable law.

4.14 Uncorrected Defects

Under Clause 46.1, the Project Manager arranges for the Client to allow access to parts of the works taken over in order to correct a defect. As stated above, if the Contractor does not correct the Defect within the Defect Correction Period the Client assesses the cost of having the Defect corrected by other people and the Contractor pays this amount. The Scope is treated as having been changed to accept the Defect.

However, under Clause 46.2, if the Contractor is not given access to correct a Defect, the Project Manager assesses the cost to the Contractor of correcting a Defect and the Contractor pays this amount. Again, the Scope is treated as having been changed to accept the Defect.

This is an unusual clause in that most contracts provide for the Contractor to have the opportunity, as well as the obligation, to correct its own defects – if it does not take the opportunity it is liable for the cost of someone else correcting it.

However in the case, particularly of high security facilities such as prisons, military establishments, research laboratories, etc., the Contractor could be denied access and be subject to the Project Manager's subjective opinion as to how much cost the Contractor would have incurred in correcting the Defect(s) if it had been allowed access and it would have to pay it.

4.15 Liability for the cost of correcting Defects

The Contractor is liable for correction of Defects, however note that under Options C, D, and E where the Contractor is reimbursed its Defined Costs, Disallowed Cost is only for the cost of:

- correcting Defects after Completion
- correcting Defects caused by the Contractor not complying with a constraint on how it is to Provide the Works stated in the Scope.

In the first bullet, the Client pays for defects to be corrected before Completion but not after Completion. There is no limit to the type or size of Defects, or why the Defect occurred, e.g. through the carelessness of the Contractor.

Whilst some may see this as the Contractor's right to repeatedly attempt to get the work right at the Client's expense it must be remembered that in the case of Options C and D these are target contracts, therefore if the Contractor is paid for correcting a Defect, not only is this bad for its reputation, but as it is not a compensation event it is not being given additional time to correct the Defect and the additional cost paid will reduce its entitlement to Contractor's share at Completion.

5   Payment

5.1 Introduction

Payment, including assessment, certification and payment of amounts due is covered by Clauses 50.1 to 53.4, and also by additional clauses within Main Options A, B, C, D, E and F.

There are separate payment provisions within the ECC, dependent on the main option chosen, each option referring to two main terms:

(i) "The Prices"

These are the various elements that make up the total price for carrying out the work and will be in the form of an activity schedule, bills of quantities, target, or some other document dependent on the main option selected.

(ii) "The Price for Work Done to Date"

This term is used in making the assessment of amounts due to the Contractor, and is again dependent on the main option chosen.

Option A	Total of the Prices for completed activities (whether in a group or not)
Option B	Quantity of work multiplied by the bill rate, proportion of lump sums
Option C	Defined Cost which the Project Manager forecasts will have been paid by the next assessment date plus the Fee
Option D	Defined Cost which the Project Manager forecasts will have been paid by the next assessment date plus the Fee
Option E	Defined Cost which the Project Manager forecasts will have been paid by the next assessment date plus the Fee
Option F	Defined Cost which the Project Manager forecasts will have been paid by the next assessment date plus the Fee

5.2 The Main Options

Option A – *Priced contract with activity schedule.*

The Prices are the lump sums for each of the activities on the Activity Schedule unless later changed in accordance with the contract (see Figure 5.1).

The Price for Work Done to Date (PWDD) is the total of the Prices for completed activities, and each completed activity which is not in a group (minor activities may sometimes be grouped together).

It is important to note that the Activity Schedule is the basis for the Contractor to be paid, it is not the Scope or a schedule of what the Contractor is intending to do in Providing the Works.

Small Industrial Unit			
Activity			
No.			
	Preliminaries		
1	Set up Site Compound		12564
2	Time based	Month 1	10875
3	preliminaries	Month 2	11200
4		Month 3	11900
5		Month 4	12600
6		Month 5	12900
7		Month 6	14500
8		Month 7	10400
9		Month 8	9200
10		Month 9	8230
11	Clear Site Compound		7231
12	Bulk Excavation	GL 1–10	86358
13		GL 10–20	71237

	Main building		
14	Reduce Level Excavation		53248
15	Ground beams	Formwork	18243
16		Reinf	19555
17		Conc	16324
18	Grd Floor Const	Formwork	21529
19		Reinf	19298
20		Conc	36589
21	Steel Frame	Delivery to site	73451
22		Erection - GL 1–6	22233
23		Erection - GL 6–12	22233
24		Erection - GL 12–18	22233
25	Roof Covering	GL 1–9	44986
26		GL 9–18	48932
27	External Cladding	GL A1–A9	32256
28		GL A9–A18	34376
29		GL A18–G18	38765
30		GL G18–G9	34372
31		GL G9–G1	32356
32		GL G1–A1	38765
33	Windows		34239
34	Internal Walls		53325
35	Joinery (1st Fix)		18987
36	M&E (1st Fix)	Mech	28253
37		Elec	31278
38	Wall Finishings		43356
39	Joinery (2nd Fix)		38972
40	Ceiling Finishes		29297
41	M&E (2nd Fix)	Mech	68795
42		Elec	72695
43	Floor Finishings		11863

Figure 5.1 (continued)

(continued)

	External Works		
44	SW Drainage		43876
45	Kerbs & Pavings		31986
46	Tarmacadam		18279
47	White Lining		9435
		Total of the Prices:	**1443575**

Figure 5.1 Sample Activity Schedule

Example

The Contractor has been awarded an Option A contract for the construction of 10 No. houses in accordance with a design produced by the Client. However, within the Activity Schedule the Contractor submitted with its tender it has not specifically included an activity for the roof covering to the houses though the roof covering is clearly detailed within the Scope.

The roof covering is deemed to be included within the Contractor's price and therefore must be assumed to have been included within the Activity Schedule. The same rule would apply to an Option C contract.

The activities on the Activity Schedule should also relate to operations on the Contractor's programme, and if, possibly due to a change of method of working, it does not, the Contractor must submit a revision to the Activity Schedule to the Project Manager for acceptance.

There will usually be more activities on the Activity Schedule than operations on the programme. This is because, for an operation which spans a number of months, the Contractor will include a number of activities.

Example

On the Contractor's programme, it has a single operation for drainage to the West elevation of the Site, which will be carried out during the months of February, March and April.

In order to optimise its use of resources, it will excavate the drain trench in February, lay the pipe in the trench in March, and backfill the trench in April, with final reinstatement of part of a road under which the drain travels to be carried out in May.

The Activity Schedule could then look like this:

Drainage to West Elevation of Site

Excavate trench	£12,200.00
Lay pipe	£8,700.00
Backfill trench	£6,600.00
Reinstate roadway	£4,800.00

The single operation on the programme will then be paid to the Contractor in four instalments.

The activities on the Activity Schedule should also relate to the Scope, if they do not, the Contractor corrects the Activity Schedule.

It is vitally important that when the Contractor submits the Activity Schedule each activity is clearly defined to prevent debates between itself and the Project Manager about what an activity is and whether it is completed.

The Price for Work Done to Date is the total of the Prices for each completed activity in the Activity Schedule, i.e. the Contractor is paid the full Price for the activity, or zero if it is not complete. The contract does not provide for incomplete activities to be paid proportionally.

A completed activity is one without Defects which would either delay or be covered by immediately following work. So, for example, if the activity was to erect a 30-metre length of pipework on and including supports and the pipework and the supports were all installed but there was a defect in one of the supports, then provided it does not delay or is covered by immediately following work, it is a completed activity, but with a Defect that needs to be corrected. If some of the supports are not yet installed, one should assess this as an activity which has not been completed.

The Contractor is not entitled to be paid for activities which are not complete at the assessment date. Assessing the Contractor's right to payment under the contract is therefore different to the traditional approach of valuing the works carried out to date.

If the effect of a compensation event is to delay the completion of an activity, the Contractor is not entitled to be paid a proportion of the price for that activity, the delayed payment having to be priced within the quotation for the compensation event that caused the delay.

Activities may include set up and clear site, time-based preliminaries and activities such as design as well as physical activities on site. It is important that the activities on the activity schedule are clearly defined not only in terms of what the work is, but where it is, so that their completion is not in doubt (see activity schedule example).

The traditional method of forecasting cash flow is by considering the programme and pricing on a month by month against the programme, inserting values against the events as they occur, allowing for delivery of materials as they are expected to arrive, and this is essentially the way that Option A works.

In practice however, particularly for a more complex element such as brickwork, whilst the labour output may be fairly constant, the Contractor may have to allow for all the material being delivered to Site prior to starting work, therefore the activity schedule may have to be "front loaded" to allow for these early deliveries, however it is important to differentiate this from front loading to create an overpayment to the Contractor.

Payment for larger or longer duration items

The same could apply for other trades such as structural steelwork and mechanical and electrical installations, where large quantities of material are delivered to Site for fixing at a later date.

As an example, an Activity Schedule which incorporates a steel frame valued at £121,000.00 could be set out as follows:

Steel frame at Subcontractor's premises awaiting delivery	£80,000.00
Delivery of steel frame to Site	£10,000.00
Erection of steel to:	
GL 1 – 4	£8,000.00
GL 4 – 8	£8,000.00
GL 8 – 12	£11,000.00
Paint steel frame	£4,000.00

It is also important on projects with preliminaries, to correctly allocate these items to the Activity Schedule, as they could contribute as much as 10%–15% of the Total of the Prices.

Payment for preliminaries

Preliminaries would normally comprise the following:

(I) FIXED PRELIMINARIES

These would comprise preliminaries which may be single items, not directly influenced by progress on Site.

Typical examples are – delivery of temporary accommodation and other equipment to Site, connection of telephones and other temporary services.

These should be identified within the Activity Schedule as single activities, e.g. "delivery of site accommodation" or a group of activities identified as "set up site". When the activities are completed, they are included within the next assessment.

(II) TIME RELATED PRELIMINARIES

These comprise preliminaries which are based on time on site rather than having any direct relationship to the quantity of work carried out.

Typical examples are – site management salaries, site accommodation hire charges, and scaffold.

These should be identified within the Activity Schedule as time-based activities, e.g. "one month hire of temporary accommodation". Again, when the activities are completed, they are included within the next assessment.

(III)   VALUE RELATED PRELIMINARIES

These would consist of preliminaries which are directly related to how much work is being carried out at any particular time.

Typical examples are – mechanical equipment, e.g. excavators, compressors, mixers.

These, which are not always referred to as preliminaries, should be included within the activities on which they are being used rather than as stand-alone activities.

Design-related activities

On Option A contracts where the Scope states that the Contractor is to design all or parts of the work, it should consider carefully how and when it requires payment for this activity.

The Contractor should not just include a single activity titled "Design" as there are many aspects of design, and one could argue that design is not completed until the project is completed or even when the Defects Certificate is issued, so the Contractor could deny itself payment for a considerable period.

Ideally, the design-related activities should be linked to various deliverables rather than generic activity descriptions such as "prepare drawings".

The various elements of design could be related to, for example, the various stages within the RIBA Plan of Work, or in its simplest form there could be separate activities for the design of various elements of the project, and for the granting of planning approval, etc.

Option B – Priced contract with bill of quantities.

The Prices are the lump sums and the amounts obtained by multiplying the rates by the quantities for the items in the Bill of Quantities.

The Price for Work Done to Date is the total of

- the quantity of work which the Contractor has completed for each item in the Bill of Quantities multiplied by the rate, and
- a proportion of each lump sum which is the proportion of the work covered by the item which the Contractor has completed.

As the quantity of work done is the actual quantity completed, Option B is essentially a remeasurement contract. However, where a change in quantity is related to a compensation event, that change is dealt with separately and the resulting amount fed back into the Bill of Quantities as a lump sum, a changed rate, or a changed quantity as appropriate.

The Bill of Quantities is a document for tendering and payment purposes; it is not Scope or Site Information, so it does not prescribe what the Contractor has to do. It is prepared by the Client, and the Contractor is assumed to have taken the Bill of Quantities as correct, and a true reflection of the work to be carried out, therefore unlike Option A, if an item is missing from the Bill of Quantities, then it is a compensation event.

Example

The Contractor has been awarded an Option B contract for the construction of 10 No. houses in accordance with a design produced by the Client. However, within the Bill of Quantities there is no item for the roof covering to the houses, though the roof covering is clearly detailed within the Scope.

Assuming the method of measurement states that roof covering is to be measured as a specific item, rather than included in another item, this would be a compensation event under Clause 60.6 in that there is a mistake in the Bill of Quantities.

Option C – Target contract with activity schedule

As with Option A, the Prices are the lump sums for each of the activities on the Activity Schedule unless later changed in accordance with the contract.

The Price for Work Done to Date is the Defined Cost which the Project Manager forecasts will have been paid by the Contractor before the next assessment date plus the Fee, so the assessment is a combination of Defined Cost paid and yet to be paid by the Contractor. For example, if the assessment date is 1 June, then costs which are forecast to be paid by 1 July are also included.

This is a change to pre-NEC3 editions of the contract which provided for the Contractor only to be paid cost that it had paid at the assessment date, the change being introduced to assist the Contractor's cash flow. Whilst the forecast may appear to be no more than a subjective guess, the Contractor has an up-to-date programme and also has probably received many of the invoices which it will be paying before the next assessment date. In practice, however, many Clients are resistant to forecasting costs a month ahead and therefore tend to amend the contract through a Z clause to retain the previous wording.

In assessing the amount due if the Contractor has paid costs in a currency different to the currency of the contract, the Contractor is paid Defined Cost in the same currency as the payments made by him. All payments are converted to the currency of the contract applying the exchange rates, in order to calculate the Fee.

Many practitioners question the role of the Activity Schedule under Option C, when unlike Option A, it is not used for assessing payments. The answer is that the Activity Schedule under Option C assists the Client in assessing tenders as they can see how the Contractor has priced its tender.

Again, as with Option A, the Activity Schedule should relate to operations on the Contractor's programme, and if due to a change of method of working it does not, the Contractor must submit a revision to the Activity Schedule to the Project Manager for acceptance.

Finalisation of Defined Cost

Under Clause 50.9 Options C, D, E and F, the Contractor is required to notify the Project Manager when a part of the Defined Cost has been finalised and provides the necessary evidence.

The Project Manager then within thirteen weeks:

* accepts that part of the Defined Cost as correct,
* notifies the Contractor that further records are needed, or
* notifies the Contractor of any errors in its assessment.

When the Contractor responds in respect of bullets 2 or 3, the Project Manager responds within four weeks. If it does not, then the Contractor's assessment is treated as correct.

Accounts and records

Under Options C, D, E and F, the Contractor is required to keep the following accounts and records (Clause 52.2) to calculate Defined Cost:

* accounts of payments of Defined Cost (under Option F, accounts of payments made to Subcontractors)
* proof that the payments have been made
* communications about and assessments of compensation events for Subcontractors, and
* other records as stated in the Scope.

The level of checking of the Contractor's accounts and records is at the Project Manager's discretion; some wish to examine each cost and its relevant back-up document, some wish only to select certain costs at random, others wish to only carry out a cursory check of costs. Some Project Managers say that if the parties really are acting in a spirit of mutual trust and co-operation as the contract requires, then there should be no need for a detailed examination of the Contractor's accounts, though this probably needs a quantum leap of faith for many Project Managers and Clients!

It is important to establish the format on which accounts and records are to be presented by the Contractor to the Project Manager, particularly in respect of assessing payments, and this can be detailed in the Scope, the following alternative methods normally being employed:

(i) The Contractor submits a copy of all of its accounts and records to the Project Manager on a regular basis.

or

(ii) The Contractor gives the Project Manager access to its accounts and records and the Project Manager takes copies of all the records it requires. Whilst not expressly stated within the contract, the Project Manager should not have to travel too far to inspect them.

It has been known for accounts and records belonging to an international Contractor to be "available for inspection" in a different country to the Site, the Contractor stating that they were available for inspection at any time during normal working hours, and in the absence of anything in the Scope, it was correct!

Again, the location and availability of accounts and records could be detailed within the Scope.

or

(iii) The Contractor gives the Project Manager direct access to its computerised accounts and records through a PIN number; the Project Manager can then take copies of the records it requires.

The Contractor's share

The Client enters the Contractor's share percentage for each share range into Contract Data Part 1, the principle being that the Contractor receives a share of any saving and pays a share of any excess when Defined Cost plus Fee is compared with the target at Completion of the whole of the works and in the final payment following the issue of the Defects Certificate.

On completion of the works, the Project Manager makes a preliminary assessment of the Contractor's share by applying the Contractors share percentage to the difference between the forecast final Price for Work Done to Date and the forecast final total of the Prices. A final assessment is made using the final Price for Work Done to Date and the final total of the Prices.

Whilst NEC practitioners often refer to the terms "pain" and "gain" in Option C and D, these terms do not actually exist within the contract. The Contractor pays or is paid the Contractor's share.

Example

At Completion of the works:

The Contractor's share percentages and the share ranges have been entered into Contract Data Part 1 as follows:

Share range	Contractor's share percentage
less than 80%	50%
from 80% to 90%	40%
from 90% to 100%	30%
greater than 100%	50%

The Total of the Prices = £3,200,000

The Price for Work Done to Date = £2,400,350

Difference = £799,650

The share range has been set in increments of 10%
10% of £3,200,200 = £320,020
From 90% to 100%
£320,000 × 30% = £96,000
From 80% to 90%
£320,000 × 40% = £128,000
Less than 80%
£159,650 × 50% = £79,825
Total Contractor's share is £96,000 + £128,000 + £79,825 = £303,825

Option D – Target contract with bill of quantities

The Total of the Prices is the total of

- the quantity of the work which the Contractor has completed for each item in the Bill of Quantities multiplied by the rate, and
- a proportion of each lump sum which is the proportion of the work covered by the item which the Contractor has completed.

so the target is based on a remeasurement of the work in accordance with the Bill of Quantities including compensation events.

Completed work is work without Defects which would either delay or be covered by immediately following work.

As with Option C, the Price for Work Done to Date (PWDD) is the Defined Cost which the Project Manager forecasts will have been paid by the Contractor before the next assessment date plus the Fee, and again the Contractor is required to keep accounts and records (Clause 52.2) to calculate Defined Cost.

Option E – Cost reimbursable contract.

The Prices are the forecast Defined Cost for the whole of the work plus the Fee.

As with Options C and D, the Price for Work Done to Date is the Defined Cost which the Project Manager forecasts will have been paid by the Contractor before the next assessment date plus the Fee.

Option F – Management contract

The Prices are the forecast Defined Cost for the whole of the work plus the Fee.

As with Option E, in terms of the Price for Work Done to Date except that the definition of Defined Cost is different from Options A, B, C, D and E and is simply the payments due to Subcontractors for work which is subcontracted, plus the costs of any work carried out by the Contractor.

The Price for Work Done to Date (PWDD) is the Defined Cost which the Project Manager forecasts will have been paid by the Contractor before the next assessment date plus the Fee.

The definition of Defined Cost is slightly different from Options A, B, C, D and E in that it is the payments due to Subcontractors for work which is subcontracted, plus the costs of any work carried out by the Contractor.

5.3 Unfixed materials on or off site

A common area of confusion is the payment to the Contractor for unfixed materials within or outside the Working Areas, i.e. materials on or off site. Unlike other forms of contract, the ECC has no express provisions for payment for such materials.

In order to be paid for these materials the Contractor should consider the payment rules for each of the main options:

Option A

As the Price for Work Done to Date is based on completed activities, the Contractor should create an activity in the activity schedule for unfixed materials.

For example, for a structural steel frame, the Contractor could include the following two activities:

1   Steel frame at Subcontractor's premises awaiting delivery   £80,000.00
2   Delivery of steel frame to Site   £10,000.00

When the complete steel frame, and not a proportion of it, is at the Subcontractor's premises awaiting delivery, the Contractor is paid the full Price of Activity 1, i.e. £80,000.

When the steel is delivered to Site, the activity is complete, the Contractor is paid the full Price of Activity 2, i.e. £10,000.

Option B

As the Price for Work Done to Date is based on bills of quantities, appropriate items may be included as method related charges.

Options C, D, E and F

As the Price for Work Done to Date is the Defined Cost which the Project Manager forecasts will have been paid by the Contractor before the next assessment date plus the Fee, the Contractor must provide accounts and records to show that it has paid for the materials or will have paid for them by the next assessment date.

Note that, in respect of unfixed materials on Site, whatever title the Contractor has to Plant and Materials passes to the Client if it has been brought within the Working Areas and passes back to the Contractor if it is removed from the Working Areas with the Project Manager's permission.

In respect of unfixed materials off Site, whatever title the Contractor has to Plant and Materials passes to the Client if the Supervisor has marked it as for this contract.

Also, in respect of marking, the contract must have identified them for payment, and the Contractor must have prepared them for marking as required by the Scope. This could include setting them aside from other stock, protection, insurance and any vesting requirements. (See also Chapter 7 regarding title to Plant and Materials outside the Working Areas.)

5.4 Assessing the amount due

The Project Manager is required to assess the amount due to the Contractor at each assessment date. The first assessment date is decided by the Project Manager to suit the Parties, and will normally be based on the time that the Contractor has been Providing the Works, the Client's procedures and timing for processing and issuing payments, and the Contractor's payment requirements and internal accounting system.

However, the first assessment must be made within the "assessment interval" after the starting date; this is normally inserted in the Contract Data as "four weeks" or "one calendar month".

Clearly, in this respect, some discussion needs to take place between the Project Manager, the Client and the Contractor in order that a mutually agreeable assessment date can be set.

Later assessment dates occur at the end of each assessment interval until

- the Supervisor issues the Defects Certificate
- the Project Manager issues a termination certificate

Note that there is no provision within the contract for a minimum certificate amount.

The Project Manager assesses the amount due at each assessment date, calculating the Price for Work Done to Date using the rules of the specific Main Option.

The Contractor submits an application for payment to the Project Manager before each assessment date, and must include details of the application, the format to be in accordance with the Scope.

In making his assessment, the Project Manager considers "an application for payment the Contractor has submitted" before the assessment date.

If the Contractor submits an application for payment, the amount due to the Contractor is:

- the Price for Work Done to Date
- plus other amounts to be paid to the Contractor (e.g. Contractor's share, value added tax, etc)
- less amounts to be paid by or retained from the Contractor (e.g. retention. delay damages).

If the Contractor does not submit an application for payment, the amount due is the lesser of:

- the amount the Project Manager assesses as due at the assessment date, as though the Contractor had submitted an application for payment before the assessment date, and
- the amount due at the previous assessment date.

When you consider these two bullet points, by using the words "the amount due is the lesser of . . . ", in effect, the Contractor probably is not due any payment unless he submits an application for payment.

Note that NEC3 and previous editions of NEC used the phrase "any application", so there was no requirement for the Contractor to submit an application. However, if the Contractor submitted an application for payment, the Project Manager considered that, if the Contractor did not, the Project Manager still made an assessment based on the requirements of the specific Main Option chosen.

This clause has in the past (and still will) caused some concern amongst Clients and Project Managers, as if the Contractor does not submit an application for payment and the Project Manager makes a mistake in making the assessment and has to correct it, interest on the correcting amount is paid. (N.B. If the correcting amount is an addition, the Client pays the interest, if a deduction, the Contractor pays the interest.)

The author believes that it should be an explicit contractual requirement for the Contractor to submit an application for payment for the assessment and payment process to take place, not for the Project Manager to have to make an assessment anyway.

It is therefore not uncommon for Clients to include within the contract the following Z clause replacing Clauses 50.2, 50.3 and 50.4:

> *"The Contractor submits an application for payment one week before each assessment date setting out the amount the Contractor considers is due at the assessment date. The Contractor's application for payment includes details of how the amount has been assessed and is in the form stated in the Scope.*
>
> *In assessing the amount due, the Project Manager considers the application for payment the Contractor submits. The Project Manager gives the Contractor details of how the amount due has been assessed.*
>
> *If the Contractor does not submit an application for payment one week before the assessment date, the Project Manager assesses the amount due as zero"*

which places the onus on the Contractor to make an application for payment, not on the Project Manager to make the assessment whether or not the Contractor has done so.

It is essential that the Contractor either submits a programme with his tender or within the timescale specified within Contract Data Part 1. Failure to do so will entitle the Project Manager to retain one quarter of the Price for Work Done to Date in his assessment of the amount due.

Note that the amount is only withheld if the Contractor has not submitted a first programme which shows the information which the contract requires, e.g. method statement, time risk allowances, etc. If the Contractor has submitted a programme which contains all the information that the contract requires, but the Project Manager disagrees with, for example, part of the method statement, or the programme has not yet been accepted, then the provision, and the associated amount retained, does not apply.

5.5 Payment

The Project Manager is required to certify payments within one week of each assessment date.

The Client pays each certified amount within three weeks of each assessment date or, if a different period is stated in the Contract Data, within the period stated. If the certificate is less than the previous one, then the Contractor pays the Client the amount due. Payments are in the currency of the contract unless otherwise stated.

Interest is paid if a payment is not made or the Project Manager does not issue a certificate which it should have issued. The interest rate is stated in Part 1 of the Contract Data and is assessed on a daily basis from the date the payment should have been made until the date when the late payment is made, calculated using the interest rate in Contract Data Part 1 compounded annually.

The interest due can be calculated on the basis of the following formula:

$$\frac{\text{Payment due} \times \text{interest rate} \times \text{the number of days late}}{365}$$

So if one assumes the following:

- Payment due = £120,000
- Payment 7 days late
- Interest rate in Contract Data Part 1 = 3%

The calculation is:

$$\frac{£120,000 \times 3\% \times 7\,\text{days}}{365}$$

Interest due = £69.04

Similarly, interest is paid on a correcting amount due to later corrections to certified amounts by the Project Manager and on interest due to a compensation event or as determined by the Adjudicator or the tribunal. The date from which interest should run is the date on which the additional payment would have been certified if there had been no dispute or mistake.

5.6 Defined Cost

Note that under Clause 52.1, Defined Cost is at open market or competitively tendered prices with deductions for all discounts, rebates and taxes which can be recovered, so the Contractor must be able to demonstrate that it has sought alternative prices.

Also, some practitioners believe that "deductions for all discounts" excludes some discounts, e.g. 2.5% cash discount, however the clause states "all discounts" so there are no exceptions.

Contractor's costs which are not included in the definition of Defined Cost are deemed to be included within the fee percentage.

The definition of Defined Cost varies according to each main option:

In Options A and B – Defined Cost is the cost of the components in the Short Schedule of Cost Components.

Options C, D and E – Defined Cost is the cost of the components in the Schedule of Cost Components less Disallowed Cost.

Option F – The Schedules of Cost Components are specifically excluded in Option F and do not apply. Defined Cost is the amount of payments due to Subcontractors for work which is subcontracted, without taking account of amounts paid or retained from the Subcontractor by the Contractor which would result in the Client paying or retaining the amount twice, and the prices for work done by the Contractor, less Disallowed Cost.

5.7 Working Areas

The Working Areas are the areas which are necessary for Providing the Works and used only for work on the particular contract.

The Working Areas are the Site, and any other area identified by the Contractor in Contract Data Part 2.

An example would be where the footprint of a building filled the site and therefore the Contractor chooses to use an adjacent piece of land for its site compound and storage of materials. In this case, it would identify the additional piece of land as an addition to the Working Areas. The adding of an area may alternatively be submitted as a proposal to the Project Manager during the carrying out of the works. It is essential that the additional area is necessary for Providing the Works and no other reason.

The significance of the Working Areas is that Defined Cost is identified in the Schedule of Cost Components and Short Schedule of Cost Components as:

1   The cost of employing people who are working within the Working Areas.
2   The cost of Equipment used within the Working Areas.
3   The cost of Plant and Materials including delivery to and from the Working Areas.
4   The cost of charges paid by the Contractor for provision and use in the Working Areas.
5   The cost of manufacture and fabrication of Plant and Materials which are designed specifically for the works and manufactured or fabricated outside the Working Areas.
6   The cost of design of the works and Equipment done outside the Working Areas.

In Options A and B, this is a consideration in pricing compensation events, in Options C, D and E it is a consideration in pricing compensation events, and also in assessing the Contractor's payments.

Any other costs not included within the Schedule of Cost Components and Short Schedule of Cost Components including overheads and also profit, are deemed to be included in the Fee.

This will include

(i)   Head office costs
(ii) Insurance and bond premiums
(iii) Profit.

(see also below)

5.8 Schedules of Cost Components

The contract contains two Schedules of Cost Components and they have two uses:

- Under Options A, B, C, D and E, they define the cost components, which are included in an assessment of compensation events.
- Under Options C, D and E, they define the cost components for which the Contractor will be directly reimbursed.

The Schedules of Cost Components do not apply to Option F (Management Contract) where Defined Cost is the amount of payments due to Subcontractors for work which is subcontracted, less Disallowed Cost.

The "Schedule of Cost Components" or the "Short Schedule of Cost Components"?

There are two Schedules of Cost Components in the contract.

This is because there are differing uses for Defined Cost, dependent on which main Option is chosen as follows:

> Options A and B: Defined Cost is only used for assessing compensation events

therefore the Short Schedule of Cost Components is used

> Options C, D and E: Defined Cost is used for assessing compensation events, and for assessing Price for Work Done to Date

therefore the Schedule of Cost Components is used, though the Short Schedule of Cost Components can be used by agreement between the Project Manager and the Contractor for assessing compensation events.

> Option F: Defined Cost is the amount of payment due to Subcontractors and the prices for work done by the Contractor itself

therefore, neither the Schedule of Cost Components or Short Schedule of Cost Components is used.

The Schedule of Cost Components

The Schedule of Cost Components only applies when Option C, D or E is used. References to the Contractor do not include its Subcontractors.

Clause 1 – People

This relates to the cost of people who are directly employed by the Contractor and whose normal place of working is within the Working Areas, i.e. the Site and any other area named in Contract Data Part 2, and also people whose normal place of working is not within the Working Areas but who are working in the Working Areas proportionate to the time they spend working within the Working Areas, i.e. people who are based off site, but who are on site temporarily.

The cost component covers the full cost of employing the people including wages and salaries paid whilst they are in the Working Areas, payments made to the people for bonuses, overtime, sickness and holiday pay, special allowances, and also payments made in relation to people for travel, subsistence, relocation, medical costs, protective clothing, meeting the requirements of the law, a vehicle and safety training.

The Schedule also relates to the cost of people who are not directly employed by the Contractor but are according to the time worked while they are in the Working Areas.

Examples would be agency labour, cleaners or security guards, in all cases paid by the hour rather than on a lump sum price basis.

The Schedule of Cost Components does not provide for people rates to be priced as part of the Contractor's tender, the principle being that the People component is real cost rather than a rate that has been forecast at tender stage months or even years before it is required to be used. This is a fairer method of establishing cost in that the risk is not the Contractor's, but in practice requires the Contractor if necessary to prove the cost of each operative, tradesman and member of staff, which can at times be laborious.

For this reason, many clients adopt a more "traditional policy" of including a schedule of rates to be priced by the Contractor at tender stage, these rates then being used for payments and compensation events where appropriate. These rates are also referred to during the tender assessment process.

Clause 2 – Equipment

This relates to the cost of Equipment (referred to as "Plant" in other contracts) used within the Working Areas. If the Equipment is used outside the Working Areas, then it is deemed to be included in the fee percentage.

If the Equipment is hired externally by the Contractor, the hire rate is multiplied by the time the Equipment is required. The cost of transport of Equipment to and from the Working Areas and any erection and dismantling costs are also separately costed.

If the Equipment is owned by the Contractor, or hired by the Contractor from a company within the parent company such as an internal "plant hire" company, the cost is assessed at open market rates (not at the rate charged by the hirer) multiplied by the time for which the Equipment is required.

Difficulty has often arisen with past editions (pre NEC3) of the contract where a piece of Equipment is owned by the Contractor, or a company within the group, in which case there would be no invoice as such to prove the cost; in many cases it was just an internal charge or cost transfer.

In those previous editions, the drafters sought to deal with the problem by establishing the weekly cost with the use of a formula:

$$\frac{\text{Purchase price of the Equipment}}{\text{Working life remaining at purchase}} \times \text{Depreciation and maintain \%}$$

Example

Purchase Price = £30,000
Working life remaining = 250 weeks
Depreciation and maintenance percentage = 20%
The weekly cost is:

$$\frac{£30,000}{250} \times 1.20 = £144.00 \text{ per week}$$

In theory, this should provide an equitable solution, but in practice it has provided some inequitable answers, and also it has to be applied to *every* piece of Equipment owned by the Contractor! One hopes it did not have too many of its own wheelbarrows on the site!

In NEC3 the drafters then changed to "open market rates multiplied by the time for which the Equipment is required" and this has followed through into NEC4.

In truth major contractors who have a large "plant" company will benefit in this respect from advantageous discount agreements with their suppliers which are well below "open market rates".

Equipment purchased for use in the contract is paid on the basis of its change in value (the difference between its purchase price and its sale price at the end of the period for which it is used) and the time related on cost charge stated in Contract Data Part 2 for the period the Equipment is required.

During the course of the contract, the Contractor is paid the time related charge per time period (per week/per month) and when the change in value is determined, a final payment is made in the next assessment.

Example

The Contractor has purchased 6 Portakabins @ £6250.00 each for use on a project of 4 years duration. This cost includes supply and delivery of the Portakabins.

- Total purchase price = 6 x £6,250.00 = £37,500.00
- On completion, the Portakabins are sold for £2,750.00 each, therefore the total sale price is £16,500.00
- The change in value is therefore £37,500.00 - £16,500.00 = £21,000.00.

Any special Equipment is paid on the basis of its entry in Contract Data Part 2.

Consumables such as fuel are also separately costed including any materials used to construct or fabricate Equipment.

The cost of transporting Equipment to and from the Working Areas and the erection, dismantling and any modifying of the Equipment is costed separately.

N.B. Any People cost such as Equipment drivers and operatives involved with erection and dismantling, or use of Equipment should be included in the cost of People, not the Equipment they work on.

Clause 3 – Plant and Materials

This deals with purchasing Plant and Materials, including delivery, providing and removing packaging and any necessary samples and tests.

The cost of disposal of Plant and Materials is credited.

Clause 4 – Charges

This covers various miscellaneous costs incurred by the Contractor such as temporary water, gas and electricity, payments to public authorities, and also payments for various other charges such as cancellation charges, buying or leasing of land, inspection certificates and facilities for visits to the Working Areas.

The cost of any consumables and equipment provided by the Contractor for the Project Manager's or Supervisor's offices is also included as direct cost.

Note that under a new provision within the NEC4 ECC, the cost of the Contractor's own consumables (referred to by the author in his many NEC training courses as "tea bags and toilet rolls"!) are not included within this cost component, and therefore unless they are in another cost component, and it is difficult to see where the costs have been relocated, they are deemed to be included within the fee percentage.

This is a change from previous editions of the ECC, where the cost is calculated by multiplying the Working Areas overheads percentage inserted by the Contractor in Contract Data Part 2 by the People cost items.

Clause 5 – Manufacture and fabrication

This relates to the components of cost of manufacture or fabrication of Plant and Materials outside the Working Areas. Hourly rates are stated in Contract Data Part 2 for the categories of employees listed.

Clause 6 – Design

This deals with the cost of design outside the Working Areas. Again, hourly rates are stated in Contract Data Part 2 for the categories of employees listed.

Clause 7 – Insurance

The cost of events for which the Contractor is required to insure and other costs to be paid to the Contractor by insurers are deducted from cost.

The Short Schedule of Cost Components

The Short Schedule of Cost Components is restricted to the assessment of compensation events under Options A and B, but if the Project Manager and Contractor agree it may be used for assessing compensation events under Options C, D and E.

Clause 1 – People

This relates to the cost of people who are directly employed by the Contractor and whose normal place of working is within the Working Areas, i.e. the Site and any other area named in Contract Data Part 2, and also people whose normal place of working is not within the Working Areas but who are working in the Working Areas proportionate to the time they spend working within the Working Areas, i.e. people who are based off site, but who are on site temporarily.

The cost component covers the full cost of employing the people including wages and salaries paid whilst they are in the Working Areas, payments made to the people for bonuses, overtime, sickness and holiday pay, special allowances, and also payments made in relation to people for travel, subsistence, relocation, medical costs, protective clothing, meeting the requirements of the law, a vehicle and safety training.

The Schedule also relates to the cost of people who are not directly employed by the Contractor but are according to the time worked while they are in the Working Areas.

Examples would be agency labour, or security guards, in both cases paid by the hour rather than on a lump sum price basis.

The amounts for People costs are calculated by multiplying the People Rates by the total time appropriate to that rate spent within the Working Areas. This is a new provision within the NEC4 ECC.

If there is no People Rate for a specific category of person, then the Project Manager and the Contractor can agree a new rate.

In the Short(er) Schedule of Cost Components in previous editions of the ECC, the Contractor would be paid similar to the Schedule of Cost Components.

Clause 2 – Equipment

This relates to the cost of Equipment used within the Working Areas.

The cost of Equipment is calculated by reference to a published list, for example BCIS (Building Cost Information Service) Schedule of Basic Plant Charges or the CECA (Civil Engineering Contractors Association) Dayworks Schedule. In Contract Data Part 2, the Contractor names the published list to be used and also inserts a percentage for adjustment against items of Equipment in the published list.

The Contractor also inserts rates into Contract Data Part 2 for Equipment not included within the published list. Any Equipment required which is not in the published list or priced within Contract Data Part 2 is then priced at competitively tendered market rates.

The time the Equipment is used is as referred to in the published list, which may have hourly, weekly or monthly rates.

By referring to the published list, whether the Equipment is then owned or hired by the Contractor is irrelevant.

The cost of transporting Equipment to and from the Working Areas and the erection and dismantling of the Equipment is costed separately. The cost of Equipment operators is included within the People costs.

Any Equipment not included within the published lists is priced at competitively tendered or open market rates.

Clause 3 – Plant and Materials

This is the same as for the Schedule of Cost Components and also deals with purchasing, delivery, providing and removing packaging and any necessary samples and tests. The cost of disposal of Plant and Materials is credited.

Clause 4 – Charges

This covers various miscellaneous costs incurred by the Contractor such as temporary water, gas and electricity payments to public authorities, and also payments for various other charges, which may or may not be relevant dependent on the project.

Clause 5 – Manufacture and fabrication

This relates to the components of cost of manufacture or fabrication of Plant and Materials outside the Working Areas. The calculation is based on amounts paid by the Contractor.

Clause 6 – Design

This is exactly the same as for the Schedule of Cost Components and deals with the cost of design outside the Working Areas. Again, hourly rates are stated in Contract Data Part 2 for the categories of employees listed.

Clause 7 – Insurance

The cost of events for which the Contractor is required to insure and other costs to be paid to the Contractor by insurers are deducted from cost.

5.9 The Fee

Whilst many practitioners believe that the Fee simply covers "overheads and profit", the definition is a little wider. All components of cost not listed in the Schedules of Cost Components are covered by the fee percentage.

There is no schedule of items covered by the fee percentage, but the following list, whilst not exhaustive, gives some examples of cost components not included in the Schedules of Cost Components.

(i)   Profit.
(ii)   The cost of offices outside the Working Areas, e.g. the Contractor's head office.
(iii) Insurance premiums.
(iv)   Performance bond costs.
(v)   Corporation tax.
(vi)   Advertising and recruitment costs.
(vii) Sureties and guarantees required for the contract.
(viii) Some indirect payments to staff.

From this one can see that there are some elements covered by the fee percentage which could be mistakenly assumed as being covered by the Schedules of Cost Components.

Prior to the NEC3 ECC, there was only one fee percentage in the contract, which was applicable to the Defined Cost.

Then the NEC3 ECC provided for the Contractor to tender two fee percentages:

(i)   The *subcontracted fee percentage* applied to the Defined Cost of subcontracted work.
(ii)   The *direct fee percentage* applied to the Defined Cost of other work.

It was essential that these two fee percentages were correctly allocated to the appropriate costs when assessing payments and compensation events, though in reality, many Contractors tended to bracket the two fee percentages together as a single fee percentage, which they were entitled to do, and this in effect made life easier for the Contractor and the Project Manager when assessing payments and compensation events.

The NEC4 ECC has now reverted to one Fee percentage in the contract, which is applicable to the Defined Cost.

5.10 Disallowed Cost

There are different clauses covering Disallowed Cost dependent on the selected main option:

Options C, D and E – Clauses C11.2.(26), D11.2.(26), E11.2.(26)

Disallowed Cost is cost which the Project Manager decides:

- is not justified by the Contractor's accounts and records

    The Contractor is obliged to keep accounts, proof of payments, communications regarding payments, and any other records as stated in the Scope (Clauses 52.2 and 52.4). If it does not have the accounts and records to prove a cost then the cost must be disallowed. It is not sufficient to merely prove that the goods or materials are on the Site for the Project Manager to see, or held off site at a designated place for inspection, as with many other contracts.
- should not have been paid to a Subcontractor or supplier in accordance with its contract

    The Contractor is obliged to submit the proposed conditions of contract for each Subcontractor to the Project Manager for acceptance, unless an NEC contract is proposed, or the Project Manager has agreed that no submission is required.

    In addition under Options C, D, E and F, the Contractor is required to submit the proposed contract data for each subcontract if an NEC contract is proposed and the Project Manager instructs the Contractor to make the submission. If the Contractor pays a Subcontractor or supplier an amount that is not in accordance with its contract with them, then this is disallowed.
- was incurred only because the Contractor did not

    – *follow an acceptance or procurement procedure stated in the Scope*

    The Scope may require the Contractor to comply with a specific procedure in respect of, for example, submission of design proposals or procurement of Subcontractors. If the Contractor does not comply with the procedure then associated costs are disallowed.

    – *give an early warning which the contract required it to give*

    If the Contractor incurs a cost that could have been avoided if the Contractor had given early warning, then it is disallowed.

> — *give notification to the Project Manager of the preparation for and conduct of an adjudication or proceedings of a tribunal between the Contractor and a Subcontractor or supplier*

and the cost of

- correcting Defects after Completion

  Correction of defects after Completion is a disallowed cost. However, correction of defects before Completion is not a disallowed cost. This is often a contentious issue with Clients and Project Managers objecting to paying the Contractor for correcting its own defects!

  However, one must consider this clause logically. The Contractor has quite possibly priced the risk of having to correct defects which are its liability within its original tender, on an Option C contract, for example, this price will be included within the target. Option C is a cost reimbursable contract until Completion, and the only way for the Contractor to be paid for this risk, should it materialise, is through the Defined Cost process, as Option C is a cost reimbursable contract until Completion. It is not a compensation event, so the target and/or the Completion Date are not changed. The Contractor is correcting the defect in its own time and potentially reducing the share it could make at Completion.

- correcting Defects caused by the Contractor not complying with a constraint on how it is to Provide the Works stated in the Scope.

  A constraint may be stated in the Scope, such as a prescribed method of working. For example, the Scope may prescribe that a hardcore sub base filling must be rolled six times with a vibrating roller of a certain weight, but the Contractor does not do so, and later a defect arises due to the Contractor's failure to comply with that stated constraint.

  If the Contractor does not comply with this constraint and a Defect occurs, then the Contractor's cost of correcting the Defect is disallowed.

- Plant and Materials not used to Provide the Works (after allowing for reasonable wastage) unless resulting from a change to the Scope.

  Excess wastage of plant or materials beyond what is considered reasonable is a Disallowed Cost. The question of what is "reasonable" can often be debatable. As with all disallowed cost it is the Project Manager's responsibility to make the decision and to disallow the cost.

- resources not used to Provide the Works (after allowing for reasonable availability and utilisation) or not taken away from the Working Areas when the Project Manager requested.

  This provision will include People and Equipment. If the Contractor does not remove Equipment when it is no longer required, then this is disallowed cost. If the Contractor is using more resources than it has planned and priced for, or its resources are inefficient, that is not a disallowed cost.

- preparation for and conduct of an adjudication, payments to a member of the Dispute Avoidance Board or proceedings of the tribunal between the Parties.

If an adjudication, the involvement of the Dispute Avoidance Board, arbitration or legal proceedings occur, then each party bears its own costs. This bullet was not introduced until NEC3 was published, so prior to NEC3 a Contractor could, in theory, refer a dispute to adjudication and whether or not it was successful, any costs arising could be included as cost and would not be disallowed.

Option F – Clause F11.2.(27)

Disallowed Cost is cost which the Project Manager decides:

- is not justified by the Contractor's accounts and records

  The Contractor is obliged to keep accounts, proof of payments, communications regarding payments, and any other records as stated in the Scope (Clauses 52.3 and 52.3). If it does not have the accounts and records to prove a cost then the cost must be disallowed. It is not sufficient to merely prove that the goods or materials are on the Site for the Project Manager to see, or held off site at a designated place for inspection, as with many other contracts.

- should not have been paid to a Subcontractor or supplier in accordance with its contract

  The Contractor is obliged to submit the proposed conditions of contract for each Subcontractor to the Project Manager for acceptance, unless an NEC contract is proposed, or the Project Manager has agreed that no submission is required.

  In addition under Options C, D, E and F, the Contractor is required to submit the proposed contract data for each subcontract if an NEC contract is proposed and the Project Manager instructs the Contractor to make the submission. If the Contractor pays a Subcontractor or supplier an amount that is not in accordance with its contract with them, then this is disallowed.

- was incurred only because the Contractor did not

  - *follow an acceptance or procurement procedure stated in the Scope*

  The Scope may require the Contractor to comply with a specific procedure in respect of, for example, submission of design proposals or procurement of Subcontractors. If the Contractor does not comply with the procedure then associated costs are disallowed.

  - *give an early warning which the contract required him to give*

  If the Contractor incurs a cost that could have been avoided if the Contractor had given early warning, then it is disallowed.

  - *give notification to the Project Manager of the preparation for and conduct of an adjudication or proceedings of a tribunal between the Contractor and a Subcontractor or supplier or*

- is a payment to a Subcontractor for

  - *work which the Contractor states that the Contractor will do itself*
    or
  - *the Contractor's management*

- and was incurred in the preparation for and conduct of an adjudication, or payments to a member of the Dispute Avoidance Board or proceedings of the tribunal between the Parties.

  If an adjudication, the involvement of the Dispute Avoidance Board, arbitration or legal proceedings occur, then each party bears its own costs. This bullet was not introduced until NEC3 was published, so prior to NEC3 a Contractor could, in theory, refer a dispute to adjudication and whether or not it was successful, any costs arising could be included as cost and would not be disallowed.

Clearly, with all the Disallowed Cost provisions, as the Project Manager is to decide that a cost is to be disallowed, then it has to be diligent enough to identify the cost, to quantify it, and to make the appropriate deductions from payments, which are not the easiest of tasks!

5.11 Project Bank Account

Note that the ECC has provision for a Project Bank Account if required, this being covered by Secondary Option Y(UK)1, details of which are discussed previously within Chapter 0.

5.12 Final assessment

The ECC does not have the equivalent of a Final Account or Final Certificate which is found in other contracts, certifying that the contract has fully and finally been complied with, and that issues such as payments, compensation events and defects have all been dealt with and, effectively, the contract can be closed.

Whilst some may say that, because of the compensation event provisions there is no Final Account, and the Defects Certificate confirms correction of any outstanding defects, there is no need for a Final Certificate, but some questions could still remain such as when, under Option E, is it too late for the Contractor to submit a cost?

However, NEC4 has introduced a new provision where, in the case of the ECC, the Project Manager makes a final assessment of amounts due to the Contractor, in effect giving closure, at least to financial aspects of the contract.

The Project Manager makes an assessment of the final amount due to the Contractor and certifies a payment, no later than:

- Four weeks after the Supervisor issues the Defects Certificate or
- Thirteen weeks after the Project Manager issues a termination certificate.

Similar to a normal payment, the Project Manager gives the Contractor details of the assessment and payment (by either Party) is made within three weeks of the assessment date or, if a different period is stated in the Contract Data, within the period stated.

If the Project Manager does not make the final assessment within the time allowed, the Contractor may issue to the Client an assessment of the final amount due. If the Client agrees with the assessment, a final payment is issued within three weeks of the assessment date or, if a different period is stated in the Contract Data, within the period stated.

The assessment of the final amount due is conclusive, unless if Option W1 is selected, a Party:

- refers a dispute about the assessment of the final amount due to the Senior Representatives within four weeks of the assessment being issued;
- refers any issues not agreed by the Senior Representatives to the Adjudicator within three weeks of the list of issues not agreed being produced, or when it should have been issued; and
- refers to the tribunal its dissatisfaction with a decision of the Adjudicator regarding the final amount due within four weeks of the decision being made.

If Option W2 is selected, a Party:

- refers a dispute about the assessment of the final amount due to the Senior Representatives or to the Adjudicator within four weeks of the assessment being issued, but it may omit this stage by virtue of W2.2(1);
- refers any issues not agreed by the Senior Representatives to the Adjudicator within three weeks of the list of issues not agreed being produced, or when it should have been issued; and
- refers to the tribunal its dissatisfaction with a decision of the Adjudicator regarding the final amount due within four weeks of the decision being made.

If the above applies under Options W1 or W2, the assessment of the final amount due is changed to include:

- any agreement the Parties reach; and
- a decision of the Adjudicator which has not been referred to the tribunal within four weeks of the decision.

If Option W3 is selected, a Party:

- refers a dispute about the assessment of the final amount due to the Dispute Avoidance Board
- refers to the tribunal its dissatisfaction with a recommendation of the Dispute Avoidance Board within four weeks of the recommendation being made.

The assessment of the final amount due is changed to include:

- any agreement the Parties reach and a decision of the Adjudicator or recommendation of the Dispute Avoidance Board which has not been referred to the tribunal within 4 week of the decision or recommendation.

A changed assessment becomes conclusive evidence of the final amount due.

6 Compensation events

6.1 Introduction

Compensation events are covered within the ECC by Clauses 60.1 to 66.3, and also by additional clauses within Main Options A, B, C, D and F.

Construction contracts have to remain flexible to the needs of end users and others and the fact that these needs may change. Compensation events within the NEC contracts are much more than about change though; they also reflect the risk profile of the contract. The Contractor may be compensated for a number of events, those events not being for reasons which are at the Contractor's own risk.

When considering the principle of compensation events under the ECC, it is appropriate first to consider how change is managed with most other standard forms of contract.

6.2 The "traditional approach" to managing change

The "traditional approach" to managing change in construction contracts is normally dealt with in three separate elements within the contract:

(i) Variation

The changes in the form of additions, omissions or substitutions are priced by measuring the changed work, then pricing the change using rates consistent with the Contractor's original tender, the pricing document for that tender normally being in the form of a bill of quantities, schedule of rates or a contract sum analysis.

Where the changed work is similar to that referred to in the pricing document, executed under similar conditions, and does not significantly change the quantity of work set out in the tender, the original rates and prices apply. Where the changed work is similar to that referred to in the tender, but not executed under similar conditions, and/or there is a significant change in the quantity of work, the original rates and prices form the basis of the pricing, i.e. they are adjusted to allow for the differences in conditions and/or quantities. There are normally provisions for the parties to agree fair rates and prices and where the work cannot be clearly defined and/or measured then daywork provisions apply.

Whilst this has always been seen as the "normal" way to price change in a construction contract, it has its drawbacks.

The first problem with the traditional approach is that the rates and prices inserted by the Contractor in its tender were calculated based on the original quantities, and whether they are right or wrong, they are being applied to changes which are outside its control. In that sense, either the Contractor or the Client may gain or lose by the changes.

The second problem is that there are rarely specific timescales within the contracts for submitting the prices and agreeing them, apart from the Final Account which is normally submitted by the Contractor and agreed by the contract administrator, months or even years after the works have been completed.

(ii) Extension of Time

Whilst the variation element deals with the changes to the prices, the effect on the completion date is dealt with as a separate process referred to as "extension of time" or "adjustment to completion date". This provision allows for the contract administrator to extend time/fix a new completion date, preventing time becoming at large and thereby postponing the Client's right to recover liquidated damages.

In that respect, the Contractor is required to submit a separate notice to the contract administrator upon it becoming apparent that a delay will occur. Normally there is a timescale within these provisions for the Contractor to submit its notice and in turn for the contract administrator to respond.

The purpose of the notice requirement is to allow the contract administrator to attempt to mitigate the problem either by stopping the delay from recurring or even omitting work in addition to making a judgement on the Contractor's entitlement to additional time.

The notice will normally identify:

(a) The cause of the delay – detail of how and why the delay is occurring or likely to occur.
(b) The estimated effect on completion – the Contractor should also give an estimate of delay in its notice so that the contract administrator can form its own opinion.
(c) The clauses on which the Contractor relies in requesting an extension of time.

(iii) Loss and Expense

Finally, contracts have provision for the Contractor to recover loss and expense in the form of prolongation costs, disruption costs, etc. which it could not have recovered under any other provision within the contract.

Loss and expense claims have from experience been given a very bad name. In reality loss and expense claims should be a process by which the Contractor can submit a well-structured, fully substantiated claim to seek to recover any loss or expense which it has incurred as a result of something which is outside its control and/or foreseeability.

Whilst the financial effect of a variation should essentially be priced with the variation, the cumulative effect of a number of variations or other factors which are not the Contractor's risk may involve the Contractor in loss and expense in having to work uneconomically, out of sequence, etc. and this provision deals with that possibility. Often agreement to loss and expense claims is a prolonged process which extends long after completion of the project. They are generally known as disruption claims.

6.3 The NEC4 approach

The NEC contracts, now NEC4, take a very different approach to pricing and managing change as contained within the compensation event provisions, whereby the event is notified, priced, assessed and implemented within a defined, and relatively short time scale, see Figure 6.1.

The contracts recognise that it is critical that the time and cost effect of a change are considered together as the result of that change, therefore the Prices, and Key Dates and the Completion Date are changed within a single provision rather than within three separate processes as with the traditional approach stated above.

A very important difference to the traditional methods is that no reference is made to any of the Contractor's original contract pricing except by the agreement of the Parties. This ensures that the Contractor's original financial and cash flow structure is unaffected by the compensation event. All time and cost effects are contained within the evaluation of the compensation event.

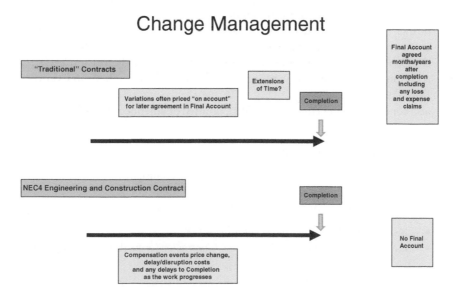

Figure 6.1 Change Management Comparison – Traditional Contracts vs NEC4 ECC

The other important difference is that the circumstances that constitute the basis of a compensation event are defined within the contract. This gives the contract drafter the facility to adjust the balance of risk between the Client and Contractor.

Experience shows that generally the requirements of the compensation events procedures are poorly understood and applied. The more onerous requirements on the Project Manager introduced in NEC3 were included to meet a particular failing as were the time bar requirements on both Parties.

Definition of a compensation event

There is no definition of a compensation event within the contract, however it could be defined as *"an event which is the Client's risk, and if it occurs, and affects the Contractor, entitles the Contractor to be compensated for any effect on the Prices, and Key Date(s) and the Completion Date"*.

The compensation events

The compensation events may be found in a number of locations within the contract:

- 21 events are stated in sub clauses 60.1(1) to (21); these are core clauses and therefore always apply, unless amended by Z clause provisions.
- Note that subclause 60.1(21) provides for additional compensation events to be stated in Contract Data Part 1.
- a further 3 events are stated in Options B and D subclauses 60.4 to 60.6; these are the options which are based on a bill of quantities prepared by the Client and therefore if, for example, there is a mistake in the bill of quantities then a compensation event applies.
- further events are included in secondary options X2 (changes in the law), X12 (partnering), X14 (advanced payment to the Contractor), X15 (limitation of the Contractor's liability for its design to reasonable skill and care) and Y(UK)2.
- additional compensation events may also be stated by the Client in Contract Data Part 1.
- also compensation events can be added, omitted or amended through Z clauses.

6.4 The compensation events

The principal compensation events are set down in Clause 60.1.

> *60.1(1) The Project Manager gives an instruction changing the Scope except*
>
> - *a change made in order to accept a Defect or*
> - *a change to the Scope provided by the Contractor for its design which is made*
>   - *at the Contractor's request or*
>   - *to comply with other Scope provided by the Client*

See Sample Project Manager's Instruction (Figure 6.2).

Contract: .... Contract No:	PROJECT MANAGER INSTRUCTION PMI No...
To: Contractor	☐ Instruction changing Scope ☐ Proposal to change Scope ☐ Other

Description

Further instructions:

☐ This instruction is/is not a compensation event   ☐ Issued as proposed change. Do not implement.

☐ Provide quotation of time/cost effect   ☐ Provide quotation of time/cost effect

☐ Start work immediately   ☐ No cost instruction

☐ Keep records

Signed: (Project Manager) ....   Date: ....

Contractor's Summary

☐ Delay to Completion is..............days/weeks

☐ No delay to Completion

☐ Change in Defined Cost plus Fee Plus/Minus £ ....

☐ No change in Defined Cost plus Fee

Note: This is a summary only. Full supporting documentation or records must accompany this statement.

Signed: (Contractor) ....   Date:...............

Copied to:

Contractor ☐ Project Manager ☐ Supervisor ☐ File ☐ Other .... ☐

Figure 6.2 Sample Project Manager's Instruction

Changes to the Scope in the ECC are normally defined in other contracts as "variations". The ECC does not use the term variations, but this is the equivalent, where the Project Manager can give an instruction changing the Scope, which can be in the form of an addition, omission or alteration to work.

An instruction can also be given in order to resolve an ambiguity or an inconsistency within the Scope. An ambiguity may be defined as "an uncertain meaning or intention" and therefore is something that is unclear; an inconsistency may be defined as a "conflict in meaning or intention" and therefore could be a mismatch or discrepancy between documents.

An example of an inconsistency is that the drawings in the Scope provided by the Client show a particular grade of concrete, but the specification shows a different grade.

Under Clause 17.1, the Project Manager or Contractor notifies the other as soon as either becomes aware of the ambiguity or inconsistency; the Project Manager then states how to resolve, it, which could include giving an instruction, and any subsequent change to the Scope is then a compensation event under Clause 60.1(1).

A compensation event which arises from an instruction to change the Scope to resolve an ambiguity or inconsistency is assessed (Clause 63.10) "as if the Prices, the Completion Date and the Key Dates were for the interpretation most favourable to the Party which did not provide the Scope". In this case it would be assessed in favour of the Contractor who is deemed to have priced the cheaper, easier or quicker alternative.

Similarly, an instruction can also be given as soon as the Project Manager becomes aware that the Scope includes an illegal or impossible requirement.

The Project Manager then gives an instruction to change the Scope to rectify the matter; again this is a compensation event under Clause 60.1(1).

Note, from the first bullet point in Clause 60.1(1), if the change to the Scope is made in order to accept a Defect (Clause 45), then it is not a compensation event.

As detailed previously in Chapter 4, the Contractor and the Project Manager may each propose to the other that the Scope should be changed so that a defect does not have to be corrected, in effect nullifying the Defect. This would not be a compensation event.

Also, from the second bullet point, a change to the Scope provided by the Contractor for its design which is made either at its request or to comply with other Scope provided by the Client is not a compensation event. So Scope provided by the Client in Part 1 of the Contract Data takes precedence over that provided by the Contractor in Part 2 of the Contract Data.

Thus it is critical that the Contractor ensures that the Scope it prepares and submits with its tender and refers to in Part 2 of the Contract Data, complies with the requirements of the Client's Scope in Part 1 of the Contract Data.

Example

The Scope included in Contract Data Part 1 states that all doors are to have brass furniture.

The Contractor states in its tender in Contract Data Part 2 that all doors will have aluminium furniture.

The requirement for brass furniture in the Client's Scope in Contract Data Part 1 will prevail.

As stated above, the Contractor notifies the Project Manager as soon as it considers that the Scope requires him to do anything which is illegal or impossible; the Parties cannot contract to do something illegal. If the Project Manager agrees it gives an instruction to change the Scope appropriately.

> *60.1(2) The Client does not allow access to and use of each part of the Site by the later of its access date and the date for access shown on the Accepted Programme.*

The Client has a fundamental obligation to allow the Contractor access to the Site (Clause 33.1). This clause recognises two dates, that shown in Contract Data Part 1, and that shown in the Accepted Programme.

The Project Manager should be aware that if the Contractor shows a date in its programme when it requires access to and use of the Site or a part of the Site and the Project Manager accepts the programme, then the Client is obliged to allow that access by the later of that date and the date in the Contract Data, failing which a compensation event occurs. So, the Contractor can relax the obligation to the date in the contract by showing a later date in the Accepted Programme, but cannot impose an earlier date than that stated in the contract.

> *60.1(3) The Client does not provide something which it is to provide by the date shown in the Accepted Programme.*

The Client is obliged to provide information, facilities, or other things as shown on the Accepted Programme. It is important that in addition to the Client stating the dates when it will provide things, the Contractor shows these requirements and the dates for providing them in its programme submitted for acceptance, so that when the Project Manager accepts the programme it accepts liability on behalf of the Client if it fails to provide it.

Note that, whilst the sub-clause refers to "the Client" it applies to anyone who acts for the Client and has an obligation to provide something, so this clause could include, for example, the Client not providing free issue materials to the Contractor, or the Project Manager not providing design information to the Contractor.

60.1(4) The Project Manager gives an instruction to stop or not to start any work or to change a Key Date.

It is important to note that the Project Manager giving an instruction to stop or not to start any work or to change a Key Date is always a compensation event, regardless of the reason why the instruction was given, however, if the Project Manager gives an instruction to the Contractor to stop a part of the work, for example, because the Contractor is working unsafely and the Contractor notifies the Project Manager that it believes it is a compensation event, then under Clause 61.4, if the Project Manager decides that the event arose from a fault of the Contractor, then it notifies the Contractor that the Price, Completion Date and the Key Dates are not to be changed.

60.1(5) The Client or Others

- *do not work within the times shown on the Accepted Programme,*
- *do not work within the conditions stated in the Scope or*
- *carry out work on the Site that is not stated in the Scope.*

The words "do not work within the times shown on the Accepted Programme" can often cause debate. What work is the Client, or Others, required to do? Also, "work within" implies a process over a period of time rather than a specific task or date so it may be difficult to determine whether someone has worked within the time shown on the Accepted Programme?

Again, it is important that in addition to the Client showing what it is going to do, the Contractor shows these requirements and the dates for providing them in its programme submitted for acceptance, so that when the Project Manager accepts the programme it accepts liability on behalf of the Client if it fails to provide it.

Note that the term "Others" applies to people who are not the Client, Project Manager, Supervisor, Adjudicator, Contractor, Subcontractor or Supplier, examples of Others being utilities companies installing or diverting electricity, gas or water services and is a traditional area of difficulty. Some Clients amend this clause to exclude their liability for "Others", thus passing the risk to the Contractor. This is unreasonable because the Contractor usually has no more contractual or financial control over the service providers than the Client.

60.1(6) The Project Manager or the Supervisor does not reply to a communication from the Contractor within the period required by the contract.

If the Project Manager or Supervisor does not reply to a communication, which could be for example, the Contractor submitting the names of a proposed Subcontractor, design proposals or a programme, etc. and not receiving a reply, then it is a compensation event as it is fundamentally a breach of contract, however if the Contractor notifies a compensation event in this respect, and the Project Manager decides that the event has no effect upon Defined Cost, Completion or meeting a Key Date, then again under Clause 61.4, it notifies the Contractor that the Price, Completion Date and the Key Dates are not to be changed.

The "period required by the contract" may either be an expressly stated period of time, for example two weeks for the Project Manager to reply to the submission of programme, or the more general "period for reply" stated in Contract Data Part 1.

> *60.1(7) The Project Manager gives an instruction for dealing with an object of value or of historical or other interest found within the Site.*

The term "object of value or of historical or other interest" could be very wide ranging and subject to interpretation, for example, what value is classed as "of value", would a mobile phone or a watch, be something of value? How old does something have to be to be classed as "historical"?

The key is that the Project Manager gives an instruction for dealing with the object, which is unlikely if a personal possession was found, which in any case would be unlikely to have an effect on the Prices and/or Completion. The clause normally deals with buried articles such as fossils, archaeological remains, but can also include munitions or other wartime remains. Such finds may also be governed by government legislation, local by-laws, etc and may involve investigation by educational establishments, museums and other authorities.

Note that under Clause 73.1, the Contractor has no title to objects of value or historical interest within the site, but notifies the Project Manager when such an object is found and the Project Manager instructs him how to deal with it.

> *60.1(8) The Project Manager or the Supervisor changes a decision which has previously communicated to the Contractor.*

The following clauses are examples of the Project Manager deciding something then changing its decision:

Clause 11.2(26) and 11.2(27) – The Project Manager decides Disallowed Cost.

Clause 30.2 – The Project Manager decides the date of Completion.

Clause 50.1 – The first assessment date is decided by the Project Manager.

Clause 61.4 – The Project Manager decides whether an event notified by the Contractor is a compensation event.

Clause 61.5 – The Project Manager decides that the Contractor did not give an early warning of the event which an experienced Contractor could have given.

This clause has been interpreted by some NEC practitioners as a provision to allow the Project Manager to change and revise its assessment of an implemented compensation event. This is an incorrect interpretation; it must be stressed that the Project Manager does not have the authority to change its assessment once it has been implemented. The Project Manager's assessment is final and the only recourse would be adjudication.

> *60.1(9) The Project Manager withholds an acceptance (other than acceptance of a quotation for acceleration or for not correcting a Defect) for a reason not stated in the contract.*

The following clauses are examples of the Project Manager having to give acceptance or non-acceptance:

Clause 16.2 – Acceptance of a proposal for adding to the Working Areas.

Clause 21.2/21.3 – Acceptance of the Contractor's design.

Clause 23.1 – Acceptance of the Contractor's design of Equipment.

Clause 24.1 – Acceptance of the Contractor's replacement key person.

Clause 26.2 – Acceptance of a proposed Subcontractor.

Clause 26.3 – Acceptance of the proposed subcontract documents for a Subcontractor.

Clause 31.3 – Acceptance of a programme.

Clause 62.3 – Acceptance of a Contractor's quotation for a compensation event.

Clause 84.1 – Acceptance of the Contractor's insurance certificate.

Options C, D, E & F

Clause 26.4 – Acceptance of the pricing information for a subcontract.

Secondary Options

X4 – accepts or does not accept a proposed alternative guarantor.

X10 – accepts or does not accept a first Information Execution Plan.

X13 – accepts or does not accept a performance bond.

X14 – accepts or does not accept a bank or insurer.

X16 – accepts or does not accept a bank or insurer.

X22 – accepts or does not accept the Contractor's forecasts of total Defined Cost.

Y(UK)1 – accepts or does not accept details of the banking arrangements.

Within each of the clauses are stated the reasons for not accepting. If the Project Manager does not accept for a different reason, or does not respond at all, then this is a compensation event.

For example, the Project Manager may not accept a proposed Subcontractor because it has never heard of them. The Contractor may offer additional information about this proposed Subcontractor, but if the Project Manager still does not accept them because it has never heard of them, then that is not that the appointment of the Subcontractor "will not allow the Contractor to Provide the Works".

60.1(10) The Supervisor instructs the Contractor to search for a Defect and no Defect is found unless the search is needed only because the Contractor gave insufficient notice of doing work obstructing a required test or inspection.

This relates to the Supervisor instructing the Contractor to search for a defect under Clause 43.1. Note that if the Contractor did not give sufficient notice of doing work obstructing a test or inspection then it is not a compensation event, so for example, if the Contractor backfilled drain trenches before the Supervisor had reasonable opportunity to inspect them, or the contract specifically stated a period for inspection and the Contractor did not allow that period, then this would be the Contractor's risk and not a compensation event.

60.1(11) A test or inspection done by the Supervisor causes unnecessary delay.

Under Clause 40.5, the Supervisor does its tests without causing unnecessary delay to the work, so it needs to act and carry out that test or inspection in a timely fashion. How long the test or inspection takes before it moves from "necessary" to "unnecessary" delay is clearly a subjective judgement, but if the Supervisor delays the Contractor beyond what would be deemed "necessary" bearing in mind the type of test or inspection, then it is a compensation event.

60.1(12) The Contractor encounters physical conditions which

- *are within the Site,*
- *are not weather conditions, and*
- *an experienced contractor would have judged at the Contract Date to have such a small chance of occurring that it would have been unreasonable for him to have allowed for them.*

Only the difference between the physical conditions encountered and those for which it would have been reasonable to have allowed is taken into account in assessing a compensation event.

It is important to interpret this final paragraph correctly, as if the Contractor did not allow anything in terms of Price and/or time within its tender for dealing with a physical condition it should only be compensated for the difference between what it found and what it *should have allowed*, so the Contractor is unlikely to be compensated for the full value of dealing with the physical condition.

Note that, under Clause 60.2, in judging the physical conditions for a compensation event, the Contractor is assumed to have taken into account:

- the Site Information – this could include site investigations, borehole data, etc.
- publicly available information referred to in the Site Information – this could include reference to public records about the Site.
- information obtainable from a visual inspection of the Site; note the use of the term "visual" – the Contractor is not assumed to have carried out an intrusive investigation of the Site.
- other information which an experienced Contractor could reasonably be expected to have or to obtain – this is a fairly subjective criterion, but precludes the Contractor from relying solely on what is contained within the Site Information. Note the clause refers to "an experienced Contractor", not "one who has a detailed local knowledge of the Site".

Under Clause 60.3, "if there is an ambiguity or inconsistency within the Site Information (including the information referred to in it), the Contractor is assumed to have taken into account the physical conditions more favourable to doing the work"; this could be the cheaper, easier or quicker alternative.

This complies with the rule of "contra proferentem" (contra = against, proferens = the one bringing forth) in that where a term or part of a contract is ambiguous or inconsistent it is construed strictly against the party which imposes or relies on it.

It is critical that the Client makes all information in its possession available to the Contractor. It cannot be selective, withholding information with a view of obtaining advantage.

It is also important to note that if the Contractor encounters unforeseen physical conditions, which are often, but not always ground conditions, it may not necessarily be compensated for the cost and time effect of dealing with it, so it must consider carefully the wording of Clause 60.1(12) and 60.2 to prove its case.

It is also important to recognise that compensation to the Contractor is assessed as the difference between what the Contractor found, and what it would have been reasonable for him to have allowed in its tender, not simply the difference between what it found and what it did allow in its tender.

60.1(13) A weather measurement is recorded

- *within a calendar month,*
- *before the Completion Date for the whole of the works, and*
- *at the place stated in the Contract Data,*

the value of which, by comparison with the weather data, is shown to occur on average less frequently than once in ten years.

Only the difference between the weather measurement and the weather which the weather data show to occur on average less frequently than once in ten years is taken into account in assessing a compensation event.

With regard to weather related delays, most contracts use the words "exceptionally adverse weather" or "exceptionally adverse climactic conditions", leaving it to the parties to determine, and in many cases to argue, what is meant by the words "exceptionally adverse". "Exceptionally" may be defined as "more than normal" whilst adverse may be defined as "unfavourable" in the sense that it impacted upon the Contractor's progress, which has to take into account the weather that would reasonably have been expected, bearing in mind the location of the site, the time of year and what the Contractor was intending to do at the time the weather condition occurred.

The ECC provides a more objective approach by referring and comparing to weather data provided by an independent party such as a meteorological office, an airport or military base.

Only the difference between the weather measurement and the weather which the weather data show to occur less frequently than once in ten years is taken into account in assessing a compensation event, so the Contractor will not be compensated for the full impact of the weather event as weather likely to occur within a ten-year period is the Contractor's risk, and it is assumed that the Contractor has already allowed for that in terms of price and programme.

Contract Data Part 1 defines the place where the weather is to be recorded. Note that Clause 60.1(13) states "at the place stated in the Contract Data", so it is vital that the historic weather data and the current weather measurements are recorded at the same place. It is important that the place chosen will truly reflect the likely conditions which will be encountered at the Site.

It also lists weather measurements for each calendar month in respect of:

- the cumulative rainfall (mm)
- the number of days with rainfall more than 5mm – whilst location and time of year is clearly a factor, from analysis of meteorological records, 5mm rainfall in a day is a fairly low level.
- the number of days in the month with minimum air temperature less than 0 degrees Celsius, and
- the number of days in the month with snow lying at a stated time GMT – there is no measure of how much snow, merely that it is lying, presumably on the ground at a stated time.

There is also provision for adding other measurements which could include wind speed, and other weather-related data. This might apply at, say, coastal or mountain locations.

Note also, that the weather measurement is recorded before the Completion Date for the whole of the works and at the place stated in the Contract Data.

As a compensation event under Clause 60.1(13) occurs when a weather measurement is recorded within a calendar month, the value of which, by comparison with the weather data, is shown to occur on average less frequently than once in ten years, there is a "trigger point" in the month at which the weather, for example the cumulative rainfall, exceeds the "once in ten years" test. Note that the

Rainfall

Date	1	2	3	4	5	6	7	8	9	10	11	12	13	14	15	16	17	18	19	20	21	22	23	24	25	26	27	28	29	30	31	mm
Daily Rainfall	2	3	2	4	2	2	3	7	8	7	2	3	2	3	2	3	4	4	2	2	3	2	3	2	3	4	2	2	2	2	2	100
Cum. Rainfall	2	5	7	11	13	15	18	25	33	41	48	50	53	55	58	62	66	68	70	73	75	78	80	83	85	88	92	94	96	98	100	
Worst Rainfall																																80

July

Period of delay

Period of delay

Compensation Event triggers on 24th

School of Thought No. 1 Compensation event triggers on 24th, only delays caused by weather after that date are assessed.

School of Thought No. 2 Compensation event triggers on 24th, consider the whole of the month versus the month falling within the "more frequently than 1 in 10 years" range.

Figure 6.3 Weather related compensation events

compensation event occurs once the trigger point has been exceeded, not once it has an effect on the progress of the works.

The weather up to the trigger point is at the Contractor's risk in terms of programme and cost and is deemed to have been included within the Contractor's tender, but once the trigger point is reached, there are two schools of thought (see Figure 6.3) as to how the compensation event operates:

(i) If the trigger point is reached on say, the 25th of the month, one must consider the whole month, but only the difference between the weather measurement and the weather which the weather data show to occur on average less frequently than once in ten years is taken into account in assessing a compensation event.

(ii) If the trigger point is reached on say, the 25th of the month, only delays incurred by the weather after the 25th are compensated for.

In either case, one must consider the worst weather in 10 years, not the average weather. Also, once a compensation event exists it is incumbent upon the Contractor to submit a quotation showing any effect that the weather had on the Prices and/or any delay to Completion and/or Key Dates. There is no automatic entitlement to be paid for the weather, just because the trigger point was exceeded.

It has to be said that assessing the effect of weather on an objective basis, rather than the traditional subjective basis, is essential, particularly as the ECC awards money as well as time, whilst most non-NEC contracts only award time, but the NEC approach presents its own challenges!

60.1(14) An event which is a Client's liability stated in these conditions of contract

The Client's liabilities are stated in Clause 80.1 but additional Client's liabilities may also be inserted into Contract Data Part 1.

There are six main categories of Client's liability in the contract and therefore covered by this clause:

(i) Liabilities relating to the Client's use or occupation of the Site.

(ii) Liabilities relating to items supplied by the Client to the Contractor until the Contractor has received them, or up to the point of handover to the Contractor.

(iii) Liabilities relating to loss or damage to the works, Plant and Materials, caused by matters outside the control of the Parties.

(iv) Liabilities arising once the Client has taken over completed work, except a defect that existed at takeover, an event which was not a Client's risk, or due to the activities of the Contractor on site after takeover.

(v) Liabilities relating to loss or wear or damage to parts of the works taken over by the Client, and any Equipment, Plant and Materials retained on Site after termination.

(vi) Any other liabilities referred to in the Contract Data. These should be stated in Contract Data Part 1.

60.1(15) The Project Manager certifies takeover of a part of the works before both Completion and the Completion Date.

Under Clause 35.2, the Client may use part of the works before Completion has been certified, in which case, it takes over the part of the works when it begins to use it, except if the use is for a reason stated in the Scope or to suit the Contractor's method of working. If takeover occurs before the Contractor has completed, and before the Completion Date, it is a compensation event.

If the Client takes over part of the works before Completion but after the Completion Date, e.g. because the Contractor is late completing and the Client wants to take over part of the works, then it is not a compensation event.

60.1(16) The Client does not provide materials, facilities and samples for tests and inspections as stated in the Scope.

Under Clause 41.2, the Contractor and the Client provide materials, facilities and samples for test and inspections in accordance with the Scope. If the Client does not comply in providing the facilities, or is late in providing them, it is a compensation event.

60.1(17) The Project Manager notifies a correction to an assumption which the Project Manager stated about a compensation event.

Under Clause 61.6, if the Project Manager decides that the effects of a compensation event are too uncertain to be forecast reasonably, it states assumptions, the quotation being based on these assumptions. If the Project Manager later notifies a correction to the assumption, it is a compensation event. In this case, the change to the Prices and any delay to Completion the Contractor had assumed in its original quotation are then replaced with corrected amounts.

Example

The Project Manager instructs the Contractor to assume the cost of £5000 for a part of a quotation for a compensation event which is too uncertain to be forecast reasonably. The cost of this element is subsequently recorded by the Contractor and found to be £6200. The Project Manager notifies a correction to the assumption, which is then priced as follows:

	£
Cost from records	6200
Project Manager's Assumption	5000
	1200
Add Fee @ 10%	120
Change to the Prices:	1320

60.1(18) A breach of contract by the Client which is not one of the other compensation events in this contract.

This is a "catch all" to cover any breaches not covered by the other compensation events; an example could be where the Client has not paid a certificate and the effect of the non-payment is that the Contractor has to delay or stop progress.

60.1(19) An event which

- *stops the Contractor completing the whole of the works or*
- *stops the Contractor completing the whole of the works by the date for planned Completion shown on the Accepted Programme,*

and which

- *neither Party could prevent,*
- *an experienced contractor would have judged at the Contract Date to have such a small chance of occurring that it would have been unreasonable to have allowed for it and*
- *is not one of the other compensation events stated in this contract*

Clause 60.1(19) is a "no fault" clause first appearing in NEC3, and dealing with an event neither Party could prevent, which stops the Contractor completing the works when planned.

It is also tied in to Clause 19.1 "Prevention", following which, if the event occurs, the Project Manager instructs the Contractor how to deal with the event.

There has been some confusion and debate regarding these clauses since NEC3 was first published.

The intention of the drafters appears to be that this would be a force majeure (superior force) clause, often referred to as an "act of God" in other contracts and legal documents, i.e. significant events such as flooding, earthquakes, volcanic eruptions and the associated dust clouds and other natural disasters which prevent the contracting parties from fulfilling their obligations under the contract, the clause essentially freeing them, or giving some relief, from their liabilities.

Whilst this appears to be what the drafters had intended, on closer examination the application of this clause is far wider.

If you take each element in turn:

The event

- *stops the Contractor completing the works or*

An event happens that prevents the Contractor from completing the works.

- *stops the Contractor completing the works by the date shown on the Accepted Programme*

An event happens that delays the Contractor completing the works. This is where the clause starts to "open up" to allow the Contractor to be compensated for less serious events.

- *Neither Party could prevent*

Clearly, if either party could have prevented the event then they would or should have. The clause refers to preventing the event, rather than taking some mitigating action to lessen its effect.

- *An experienced Contractor would have judged at the Contract Date to have such a small chance of occurring that it would have been unreasonable for him to have allowed for it*

At tender stage, the likelihood of the event was so low, the Contractor would not have allowed for it.

- *Is not one of the other compensation events stated in this contract*

If the matter could have been dealt with as one of the other compensation events then it should be. Clause 60.1(19) then clearly deals with the exceptional event.

Examples of where this clause would at least cause confusion:

Example 1

Clause 60.1(13) has been deleted from the contract by means of a Z clause.

One would then assume that the risk of unforeseen weather conditions is the Contractor's and it should have allowed for it within its tender.

The site is affected by a severe storm which occurs less frequently than once in 50 years, which delays completion of the works by one week.

The Contractor states that:

The weather stopped him completing the works by the date shown on the Accepted Programme. Assume that the Contractor can prove that.

Neither party could prevent the event. Whilst one can take measures to lessen the impact of a storm, one cannot prevent the storm from occurring.

An experienced Contractor would have judged at the Contract Date to have such a small chance of occurring that it would have been unreasonable to have allowed for it. The Contractor states that it could not have predicted or allowed for such a rare event and its consequences.

It is not one of the other compensation events. If Clause 60.1(13) has been deleted from the contract by means of a Z clause, one could argue that it is no longer one of the other compensation events.

Example 2

Could Clause 60.1(19) also extend to loss or damage occasioned by insurable risks such as fire, lightning, explosion, storm, flood, earthquake, terrorism? Although both Schedules of Cost Components require that the cost of events for which the contract requires the Contractor to insure should be deducted from cost, the Contractor could still recover any delay to Completion through the compensation event.

If Clause 60.1(19) is intended as a form of force majeure provision, rather than including a fairly long-winded definition within the clause, why not just state:

"A Force Majeure event occurs"

and allow the parties to interpret this internationally recognised term as they do in other contracts worldwide.

60.1(20) The Project Manager notifies the Contractor that a quotation for a proposed instruction is not accepted.

This new compensation event is related to Clause 65.3, where the Contractor has submitted a quotation for a proposed instruction, and the Project Manager has replied to the quotation stating that the quotation is not accepted, and that the Project Manager may issue an instruction, notify the instruction as a compensation event, and instruct the Contractor to submit a quotation for additional work that may not go ahead.

It gives the Contractor some recompense for preparing and submitting quotation for the proposed instruction.

60.1(21) Additional compensation events stated in Contract Data Part 1.

This new compensation event enables any additional compensation events to be detailed within the Contract Data rather than as Z clauses.

Many users delete compensation events, for example:

- 60.1(12) especially in a design and build contract. If this is deleted, also delete Clause 60.2.
- 60.1(13) weather events.

Clause 19.1 and 60.1(19) are connected. As discussed above regarding Clause 60.1(19), it is recommended that it be deleted or at least reworded to create a force majeure provision.

Delete:

- Clause 19.1
- Clause 60.1(19)

As stated above, there are further compensation events within the Secondary Options.

6.5 Additional compensation events within Options B and D

Under Options B and D, the options which use Bills of Quantities, there are three additional compensation events:

Clause 60.4

A difference between the final total quantity of work done and the quantity for an item in the Bill of Quantities is not a compensation event in itself, unless it satisfies *all three* bullet points within the clause.

This clause has caused some confusion, and has even led practitioners to delete it as a Z clause because it is seen either as confusing or unfair in its application, but if we examine each of the bullet points in turn:

- *the difference does not result from a change to the Scope,*
    If it had resulted from a change to the Scope, then it would have been dealt with as a compensation event under Clause 60.1(1).
- *the difference causes the Defined Cost per unit of quantity to change,*
    The purchase price of materials may be affected by a change in the quantity, for example a price reduction or increase may be applicable when the number of units exceeds or falls below a certain amount, or the cost of mobilisation or delivery could have changed as the same vehicle delivers fewer units.
- *the rate in the Bill of Quantities for the item multiplied by the final total quantity of work done is more than 0.5% of the Total of the Prices at the Contract Date.*

This final bullet point compares the final extended price for that item in the bills of quantities with the Total of the Prices at the Contract Date (many contracts refer to this as the "Tender Sum" or the "Contract Sum".)

If the extended price is more than 0.5% of the Total of the Prices at the Contract Date, and both of the previous bullet points are satisfied, then it is a compensation event. This final bullet excludes items of a "minor value" in the Bill of Quantities.

Example

Original quantity of work

Road Gullies   150 No. @ £125.00   £18,750.00

Due to an error in measuring the road gullies when preparing the bill of quantities, the final quantity is only 120 No., so:

Final quantity of work

Road Gullies   120 No. @ £125.00   £15,000.00

The Total of the Prices at the Contract Date is £2,000,000.

0.5% × £2,000,000 = £10,000.00, therefore, assuming that all three conditions are satisfied, then it is a compensation event.

Clause 60.5

A difference between the final total quantity of work done and the quantity for an item on the Bill of Quantities which delays Completion or the meeting of a Condition stated for a Key Date.

Again, a difference in quantity alone is not a compensation event, but if the difference also delays Completion of the whole of the project or the meeting of a Key Date then it is a compensation event.

Clause 60.6

This clause covers corrections of mistakes in the Bill of Quantities which are departures from the rules for measurement, or are due to ambiguities or inconsistencies.

It is the Client's responsibility to properly include and describe items in the Bills of Quantities. If a mistake arises, the Project Manager gives an instruction correcting the mistake, and it is dealt with as a compensation event. This will include items included within the Scope which are not included within the Bills of Quantities.

Finally, under Clause 60.7, where there is an inconsistency between the Bill of Quantities and another document, e.g. the Scope, the Contractor is assumed to have taken the Bill of Quantities as correct.

Whilst the Scope defines the work the Contractor has to do, the Bill of Quantities is the document on which it prices its tender and therefore it can safely assume that the Bill of Quantities correctly represents the work it has to price. Again, this reinforces the Client's responsibility to properly include and correctly describe items in the Bills of Quantities.

6.6 Notifying Compensation Events

Compensation events may either be notified by the Project Manager, or the Contractor can notify a compensation event to the Project Manager (see Figure 6.4).

(i) The Project Manager notifies the Contractor of the compensation event at the time of giving the instruction, issuing a certificate, or changing an earlier decision (Clause 61.1). It also instructs the *Contractor* to submit quotations, unless the event arises from a fault of the *Contractor* or the event has no effect upon Defined Cost, Completion or meeting a Key Date. The *Contractor* puts the instruction or changed decision into effect.

(ii) The Contractor can notify a compensation event, but must do so within eight weeks of becoming aware of the event, otherwise it is not entitled to a change in the Prices, the Completion Date or a Key Date, unless the event arises from the Project Manager or the Supervisor giving the instruction, issuing a certificate, or changing an earlier decision. In other words, the Project Manager should have notified the event to the Contractor but did not (Clause 61.3).

The intention of (ii) is to compel the Contractor to notify compensation events promptly, otherwise any entitlement to additional time and money is lost. Note that the eight-week rule requires the Contractor to notify the compensation event to the Project Manager within eight weeks of becoming aware of it, not just to have given early warning or mentioned the possibility of a compensation event in a discussion with the Project Manager.

The clause therefore covers compensation events initiated by the Contractor, rather than the Project Manager, examples being the weather, unforeseen ground conditions, which the Project Manager may not be aware of unless the Contractor had notified it.

An example of the Project Manager failing to notify a compensation event when it should have, would be if an instruction was issued by the Project Manager to the Contractor changing the Scope, but at the time the Project Manager did not notify the compensation event, and the Contractor in turn did not notify

Contract:	COMPENSATION EVENT NOTIFICATION
Contract No:	CEN No...

To: Contractor/Project Manager

You are notified of the following compensation event:

Description

The Compensation Event is likely to have an effect on:

☐ Price
☐ Completion

Has an early warning been given?

☐ Yes
☐ No

Quotation Required?

☐ Yes
☐ No

Signed: Project Manager/Contractor):............................	Date: ....

Copied to:

Contractor ☐   Project Manager ☐   Supervisor ☐   File ☐   Other .... ☐

Figure 6.4 Compensation Event Notification

either. Even if the compensation event is not notified within eight weeks, the responsibility remains with the Project Manager as it gave the instruction and should have notified the event to the Contractor but did not.

Various opinions have been published about the enforceability and effectiveness of time bars in contracts such as NEC3, commentators particularly debating whether the clause is a condition precedent to the Contractor being able to recover time and money, and whether a party, the Client can benefit from its own breach of contract to the detriment of the injured party, the Contractor.

For example, if the Client does not provide something which it is to provide by the date for providing it shown on the Accepted Programme, this is a valid compensation event under Clause 60.1(3), but can the Client prevent the Contractor from receiving any remedy because the Contractor failed to notify the Project Manager within the eight-week limit, and possibly, if Completion is delayed does the Contractor have to pay delay damages?

The author, whilst not being a lawyer, is of the opinion that the parties are clear as to what the terms of their agreement are at the time the contract is formed; it is also clear what happens if the Contractor does not notify a compensation event within the time stated within Clause 61.3, as it is protected against notifications which the Project Manager should have given but did not, and therefore the time bar must be effective and enforceable.

As there are likely to be several compensation events during the period of the contract, a schedule of compensation events should be kept, identifying and numbering each, with additional information about each one, see Figure 6.5.

The author has found the schedule useful for tracking compensation events as the project progresses, but also at the end of each project to review the schedule and each compensation event notified during the life of the project as "lessons learned" to consider in drafting Scopes for future contracts.

Clause 61.4 covers three possible outcomes to the Contractor's notification of a compensation event under Clause 61.3.

Negative reply

If the Project Manager responds by stating that an event notified by the Contractor

- arises from a fault of the Contractor,
- has not happened and is not expected to happen,
- has not notified within the timescales set out in these conditions of contract,
- has no effect upon Defined Cost, Completion or meeting a Key Date, or
- is not one of the compensation events stated in the contract,

the Project Manager notifies the Contractor of its decision that the Prices, Completion Date and the Key Dates are not to be changed and states the reason. Note that the Project Manager only needs to name one of these as its reason for refusing a compensation event.

SCHEDULE OF COMPENSATION EVENTS

Contract: ....

Contract No: ....

Ref.	Date notified	Origin? Project Manager or Contractor	Description of compensation event	Relevant Clauses	Action with PM or Contractor	Date Concluded	Notes
1							
2							
3							
4							
5							
6							
7							
8							
9							
10							

Figure 6.5 Sample Schedule of Compensation Events

Positive reply

If the Project Manager decides otherwise, it notifies the Contractor that it is a compensation event, and instructs him to submit quotations.

No reply

If the Project Manager does not reply within the time allowed, then the Contractor may notify the Project Manager to that effect. If the Project Manager does not reply within two weeks of the Contractor's notification, then it is deemed acceptance that the event is a compensation event and an instruction to submit quotations. It is a condition precedent upon the Contractor that the notification be given before deemed acceptance can occur.

6.7 Failure to give early warning

As previously stated in Chapter 3, under Clause 61.5 and 63.7, if the Project Manager decides that the Contractor did not give an early warning of the event which an experienced contractor could have given, it notifies this to the Contractor when it instructs it to submit quotations. If the Project Manager has done so, the event is assessed as if the Contractor had given early warning.

Failure to give an early warning can also be considered as Disallowed Cost under Options C, D, E and F, Disallowed Cost being cost which the Project Manager decides was incurred only because the Contractor did not give an early warning which the contract required him to give.

6.8 Project Manager's Assumptions

When the Contractor is instructed to submit a quotation, there may be a part of the quotation which is too uncertain to be forecast reasonably. In this case the Project Manager should state what the Contractor should assume, which could be a cost and/or time effect (Clause 61.6).

Subsequently, when the effects are known or it is possible to forecast reasonably it notifies a correction to the assumption and it is dealt with as a correction under Clause 60.1(17) (See example under Clause 60.1(17) above). Note, the Project Manager must state the assumption; if the Contractor makes assumptions when pricing the compensation event, then they are not corrected.

Example

The Contractor is instructed by the Project Manager to excavate a trial pit to establish the nature of the sub strata in a part of the Site, so that the design for a new foundation for an oil storage tank can be finalised. The Project Manager tells the Contractor that it and the structural engineer will inspect the sides to the excavation as the Contractor progresses with the work and will instruct the Contractor when to stop excavating.

As the plan size and depth of the excavation is uncertain, the Contractor is unable to make a reasonable forecast of proposed changes to the Prices and any delay to Completion, so the Project Manager states that the Contractor should allow an excavation 4 metres × 4 metres and to a depth of 3 metres, and to backfill the excavation on completion when instructed. The Contractor submits its quotation which is accepted by the Project Manager, and the compensation event is then implemented.

Subsequently, the excavation is actually required to be 4 metres × 4 metres and to a depth of 4.5 metres. The Project Manager therefore notifies a correction to an assumption, which is a compensation event under Clause 60.1(17) and instructs the Contractor to submit a quotation. The resulting quotation forecasts the Defined Cost of the original trial pit, forecasts the Defined Cost of the actual trial pit and the fee percentage is then added to the difference.

Note that no compensation event is notified after the issue of the Defects Certificate.

6.9 Alternative quotations

Under Clause 62.1, the Project Manager may require the Contractor to provide alternative quotations, for example to carry out the work using existing resources or, alternatively, to provide additional resources which might mitigate delay. Discussion will normally take place between the Project Manager and the Contractor to ascertain what is feasible and practicable.

6.10 Contractor's quotations

Note that under Clause 62.2 within the Contractor's quotation it needs to consider proposed changes to the Prices, and any delay to the Completion Date and/or Key Dates.

The quotation must include for any risk allowances for cost and time that have a significant chance of occurring, and are not compensation events. Obviously, if a risk does materialise and it is a compensation event, for example the finding of unforeseen physical conditions or bad weather, then that would have to be dealt with as a separate compensation event.

Changes to the Prices

The changes to the Prices are assessed as the effect of the compensation event upon:

- the actual Defined Cost of work done by the dividing date
- the forecast Defined Cost of the work yet to be done by the dividing date, and
- the resulting Fee.

Assessment of compensation events is based on their effect on Defined Cost and time, with Defined Cost as defined within the Schedule of Cost Components or the Short Schedule of Cost Components.

This is different from most standard forms of contract where variations are valued using the rates and prices in the contract as a basis. The reason for this policy within the ECC is that no compensation event which is the subject of a quotation is due to the fault of the Contractor or relates to a matter which is at its risk under the contract. It is therefore appropriate to reimburse the Contractor its forecast additional costs or Defined additional costs arising from the compensation event. Figures 6.6 and 6.7 give an example of the calculation of a quotation for a compensation event, and the submission of the quotation to the Project Manager, respectively.

Example

Assessing compensation events

On an Option C contract, the Project Manager gives an instruction to the Contractor as follows:

"Construct an additional 50 metres long × 150mm diameter drainage run as shown on Drawing No. 32."

The Contractor's quotation will comprise changes to the Prices and also any delay to the Completion Date and any Key Dates in the contract.

Delay to Completion

First the Contractor considers whether there will be any delay to planned Completion. The Contractor should provide details to demonstrate to the Project Manager the basis on which the delay has been calculated; this will probably require him to issue a programme, or an extract from the programme with its quotation.

It assesses that this additional work will take him two weeks to complete and as the instruction has been issued only three weeks before planned Completion, by the time it has obtained the materials to carry out the work it will delay Completion by one week.

Delay to completion has then been assessed as one week.

Changes to the Prices

First, it must be remembered that the change to the Prices is not assessed by using the Contractor's rates from the activity schedule. It must be assessed as the effect of the compensation event upon the Defined Cost of the work and the resulting fee.

```
┌─────────────────────────────────────────────────────────────────────┐
│            COMPENSATION EVENT QUOTATION EXAMPLE                       │
│          (BASED ON SCHEDULE OF COST COMPONENTS)                       │
├─────────────────────────────────────────────────────────────────────┤
│              ADDITIONAL DRAINAGE (PMI No. 83)                         │
├─────────────────────────────────────────────────────────────────────┤
```

PEOPLE 1

Labourers

2 labourers x 2 weeks x £320 per week   1280.00

Ganger (part time)

1 ganger x 2 weeks x 20% x £400 per week   160.00

JCB Operator

Note that the JCB driver is priced
with the "People" not the "Equipment".

JCB Driver

1 week x £480 per week   480.00

Contractor asseses 1 week delay to planned Completion

As Completion is delayed, the Contractor must also
consider People costs in respect of the extended time
within the Working Areas.

Supervision

Site Manager

1 week x £650 per week   650.00

Section Foreman

1 week x £500 per week   500.00

                          Total People Cost: 3070.00

Figure 6.6 (continued)

(continued)

EQUIPMENT 2

Excavator x 1 week x £1050 per week	1050.00

Fuel – 300 litres per week x 1.40 per litre   420.00
(11 litres per hour when working)

*Again, as Completion is delayed, the Contractor must
consider Equipment costs in respect of the extended
time within the Working Areas*

Temporary Office Hire
3 x 1 week x £150 per week   450.00

Temporary Toilet Hire
2 x 1 week x £170 per week   340.00

                     Total Equipment Cost: 2260.00

PLANT AND MATERIALS 3

55 metres x 150mm diameter pipe x 14.20 (nett cost)   781.00
per metre.
(50 metres + 10% waste)

                Total Plant and Materials Cost: 781.00

Total Forecast Defined Cost :   6111.00

              Fee percentage (from Contract Data Part 2) 10%   611.10

                          Total of Quotation: 6722.10

Figure 6.6 Example of Calculation Compensation Event Quotation

Contract:	Compensation Event Quotation
Contract No:	CE No..

To: Project Manager

Summary of Quotation

Change to Prices:

Change to Completion Date:

Signed: .... (Contractor)	Date: ....

The Project Manager's reply is:

☐ An Instruction to submit a revised quotation
☐ Acceptance of the quotation
☐ A notification that a proposed instruction will not be given
☐ A notification that the Project Manager will be making his own assessment

Signed: .... (Project Manager)	Date: ....

Copied to:

Contractor ☐   Project Manager ☐   Supervisor ☐   File ☐   Other .... ☐

Figure 6.7 Example of Compensation Event Quotation Submission

It is worth some clarification here regarding the term "dividing date". Clause 63.1 in previous editions of the ECC stated:

- the actual Defined Cost of the work already done,
- the forecast Defined Cost of the work not yet done, and
- the resulting Fee.

This led to some confusion as to when, in preparing his quotation, the Contractor changed from "actual Defined Cost" to "forecast Defined Cost", so the new term "dividing date" was included within the NEC4 ECC.

Again, as previously stated above, for a compensation event arising from the Project Manager or Supervisor giving an instruction or notification, issuing a certificate or changing an earlier decision, the "dividing date" is the date of that communication.

For other compensation events, the "dividing date" is the date of the notification of the compensation event.

So, in preparing his quotation, the Contractor uses actual Defined Cost of work which has been done by the dividing date, and forecast Defined Cost of the work yet to be done by the dividing date.

Note that if the effect of a compensation event is to reduce the Prices, the Prices are not reduced unless it is a change to the Scope (Clause 60.1) or a correction to an assumption (Clause 60.1(17).

Once the Project Manager has instructed the Contractor to submit a quotation, the Contractor has a maximum of three weeks in which to submit its quotation to the Project Manager, although the period may be extended by agreement between the Project Manager and the Contractor under Clause 62.5 (see Figure 6.8).

Contractor submits a quotation	Project Manager replies to quotation
3 weeks	2 weeks
Time for the Contractor to submit a quotation may be extended by agreement with the Project Manager (Clause 62.5)	Time for the Project Manager may be extended by agreement with the Contractor (Clause 62.5)
If the Project Manager does not agree to extend the time for submission, he notifies the Contractor that he will be making his own assessment. (Clause 62.3)	If the Project Manager does not reply within 2 weeks the Contractor may notify the Project Manager to that effect. If the Project Manager does not reply to the notification within 2 weeks, the Contractor's notification is treated as acceptance by the Project Manager. (Clause 62.6)

Figure 6.8 Timing for Submission and Assessment of Quotations

If multiple compensation events arise, the quotations should be carried out sequentially, separately and prospectively.

However, when a number of events arise over a short period of time, it may be sensible, by agreement, to consider them together. The impact of other compensation events existing prior to the dividing date should be recognised before making the new compensation event assessment.

If the Project Manager and Contractor do not agree to extend the time, once the three weeks have expired the Project Manager notifies the Contractor that it will be making its own assessment (Clause 62.3). If the Contractor has not submitted a quotation, the Project Manager notifies within two weeks of the date the Contractor should have submitted it. The Project Manager only has to notify the Contractor within the two-week period that it will making its own assessment, and the details of the assessment must be notified to the Contractor within the time period the Contractor had to submit the quotation, i.e. normally within a further three weeks.

Once the Contractor has submitted its quotation to the Project Manager, the Project Manager is required to reply as follows:

- it could accept the quotation, in which case it is implemented;
- it could instruct the Contractor to submit a revised quotation, in which case it explains its reasons for doing so to the Contractor. The Contractor then has a further three weeks to re-submit it, though in most cases it will review and re-submit within a shorter timescale;
- it could notify the Contractor that it will be making its own assessment, in which case the Project Manager has up to three weeks to make that assessment from the time of the notification.

If the Project Manager makes an assessment and the Contractor disagrees, then the Contractor can refer the matter to Adjudication. There is no contractual provision in NEC4 to "discuss/negotiate" the Project Manager's assessment.

Delays to the Completion Date and/or Key Dates

Note the clause states "any delay" not "any effect", so the Completion Date and any Key Dates can only be changed to a later date or remain unchanged due to a compensation event.

In assessing any delay to the Completion Date, the delay is the length of time that, due to the compensation event, the planned Completion is later than planned Completion as shown on the Accepted Programme (Clause 63.5), so any terminal float attached to the whole programme is retained by the Contractor. Similarly, any delay to a Key Date is the length of time that, due to the compensation event, the planned date when the Condition will be met is later than the planned date as shown on the Accepted Programme.

In preparing his quotation, in terms of any delay to Completion and/or Key Dates, the Contractor determines the effect that the compensation event, and no other event has upon Completion and/or Key Dates, and uses the Accepted Programme current at the dividing date in order to do that.

For a compensation event arising from the Project Manager or Supervisor giving an instruction or notification, issuing a certificate or changing an earlier decision, the "dividing date" is the date of that communication.

For other compensation events, the "dividing date" is the date of the notification of the compensation event.

To help determine any delays, the Contractor should first recognise the impact of:

- alterations to the Accepted Programme resulting from other compensation events occurring prior to the dividing date,
- delays to planned completion to operations resulting from other causes of delay occurring prior to the dividing date and which are not compensation events, and
- better or worse actual progress for operations that have started or should have started prior to the dividing date.

From these provisions one can see that no compensation event can result in an earlier Completion Date, though it may give rise to an earlier planned Completion.

It is worth noting that the Contractor will usually show as little float as possible in its Accepted Programme, i.e. all operations are close to or on the critical path, so that any compensation event delays Completion and in turn the Completion Date. This has been highlighted in Chapter 3 as one of the reasons why it is important for the Project Manager to spend time gaining a full understanding of the Contractor's programme and properly evaluating it before accepting it.

No reply to a quotation

There has always been concern amongst Contractors who were working under the 1st and 2nd Editions of the ECC who queried what remedy was available if the Project Manager failed to reply to a quotation within the period required by the contract, the simple answer being that failure to reply to a communication within the period required by the contract is another compensation event under Clause 60.1(6), which does not remedy the problem with the initial compensation event.

The drafters of NEC3 considered the problem and introduced a new Clause 62.6, which, in seeking to provide a remedy to the Contractor for the Project Manager's failure, seems to favour the Project Manager.

If the Project Manager does not reply to a quotation within the time allowed, the Contractor may notify him to that effect. If the Contractor has submitted more than one quotation it states in that notification which one it proposes is to be accepted. If the Project Manager does not reply within two weeks of the notification, the notification from the Contractor is treated as acceptance of the Contractor's quotation by the Project Manager (Clause 62.6).

It is somewhat concerning that a pre-condition to deemed acceptance is that the Contractor must issue a notification giving the Project Manager a further, but final two weeks in which to deliberate, yet if the Contractor fails to provide a quotation within the time allowed, the Project Manager's notification tells the Contractor that it will be making its own assessment, so there is no further time for the Contractor to respond.

In addition, Secondary Option W1 (Dispute Resolution) specifically allows the Client to refer a dispute about a quotation for a compensation event which is treated as having been accepted, to be referred to the Adjudicator, so the Contractor still does not have final acceptance! This is a curious inclusion when Clause 62.6 states that no reply to the Contractor's notification is treated as acceptance by the Project Manager. One would assume that an adjudicator whose role is to enforce the contract would have to find in favour of the Contractor?

6.11 Changes to the Scope that omit work

When the Project Manager gives an instruction changing the Scope by omitting work, the method of changing the Prices and dealing with the Completion Date is often misunderstood.

The tendency is for parties to simply delete or even to just ignore the item in the pricing document (activity schedule or bill of quantities), for example in an Option A contract, the activity in the activity schedule which represents the omitted work is simply reduced or not paid, or in an Option B contract the quantity in the bill of quantities is reduced, or if a single item in the bill, the item is just not paid. Admittedly, this is usually the correct way to price change in a non-NEC contract, but in an NEC contract, both assumptions are incorrect.

Omitted work due to a change to the Scope is assessed not by omitting or remeasuring the item in the relevant pricing document, but by forecasting the Defined Cost of that omitted work and adding the fee percentage. The resulting amount (Change to the Prices) is then adjusted against the value in the pricing document.

It must be remembered that the principle with compensation events in terms of their financial value is that the Contractor neither gains or loses as a result of the compensation event; the Contractor is compensated so that it is in the same financial position after the event as it would have been before the event.

If a Contractor has included low rates in its tender for work which is subsequently omitted it is quite probable that the application of forecast defined Cost plus Fee will give rise to a negative value; this is testament to the fact that the Contractor would have made a loss if the work had not been omitted. Similarly if high rates exist, the Contractor will retain the margin it would have made if it had carried out the work.

Example

In an Option A contract, the Activity Schedule includes an activity entitled "boundary fence", the activity having a price of £26,000. The Project Manager gives an instruction to omit the boundary fence; this is a change to the Scope and therefore a compensation event under Clause 60.1(1).

The tendency is to assume that the Contractor is simply not paid for that activity as it will not be carried out, however the correct way to assess the change to the Prices is to forecast the Defined Cost of the work not yet done and add the resulting Fee.

Let us assume that the Contractor has already placed an order in the sum of £20,100 with a fencing subcontractor, and can therefore prove what its costs would have been had the work not been omitted and the fee percentage is 8%.

The Forecast Defined Cost is then the value of the subcontract order:

Subcontract Order to supply and install fencing	£20,100.00
Add 8%	£1,608.00
Change to the Prices	£21,708.00

The amount of the quotation, assuming it has been accepted by the Project Manager is then used to change the Prices, so either a new activity is then inserted to the value of - £21,708.00 or the activity priced at £26,000.00 is omitted and replaced with an activity priced at £26,000.00 – £21,708.00 = £4,292.00.

The Contractor in this case retains the profit it would have made if the boundary fence had not been omitted. Clearly, if the change to the Prices had been more than £26,000 the Contractor would retain the loss.

The same principle would apply with all the main options, for example if it was an Option B contract and the boundary fence was an item in the Bill of Quantities, then the Bill of Quantities would again be changed in the same way, and again the Contractor retains the profit (or loss) it would have made if the boundary fence had not been omitted.

6.12 The Project Manager's assessments

Under Clause 64.1, the Project Manager may, for the following reasons, assess a compensation event:

- if the Contractor has not submitted a required quotation and details of its assessment within the time allowed;

- if the Project Manager decides that the Contractor has not assessed the compensation event correctly in a quotation and it does not instruct the Contractor to submit a revised quotation;
- if, when the Contractor submits quotations for a compensation event, the Contractor has not submitted a programme or alterations to a programme which the contract requires him to submit; or
- if, when the Contractor submits quotations for a compensation event, the Project Manager has not accepted the Contractor's latest programme for one of the reasons stated in the Contract.

These are all derived from some failure of the Contractor either to submit a quotation, to assess the compensation event correctly or to submit an acceptable programme.

If the Project Manager makes its own assessment it should put itself in the position of the Contractor, giving a properly reasoned assessment of the effect of the compensation event, detailing the basis of its calculations and providing the Contractor with details of that assessment. A Project Manager's assessment is not simply the quotation returned to the Contractor with "red pen" reductions down to a figure the Project Manager is prepared to accept.

It is also important to recognise that the Project Manager is not sending its assessment to the Contractor for its acceptance; it is the final decision under the contract, the Contractor's only remedy being adjudication.

Under Clause 64.2, the Project Manager assesses the programme for the remaining work and uses it in the assessment of a compensation event if:

- there is no Accepted Programme,
- the Contractor has not submitted a programme or alterations to a programme as required by the contract, or
- the Project Manager has not accepted the Contractor's latest programme for one of the reasons stated in the contract.

The Project Manager notifies the Contractor of its assessment and gives details within the period allowed for the Contractor's submission of its quotation for the same compensation event.

Under Clause 64.4, if the Project Manager does not assess a compensation event within the time allowed the Contractor may notify him to that effect. If the Project Manager does not reply to the notification within two weeks, the Contractor's notification is treated as acceptance of the quotation by the Project Manager.

The only remedy the Project Manager has, if it later finds the quotation is not acceptable is to refer it to adjudication. However it must be remembered that the role of the Adjudicator and the adjudication process is to enforce the contract, therefore if the Contractor's quotation has been submitted in accordance with the contract, then the Client will probably be unsuccessful in the adjudication.

6.13 Proposed instructions

Under Clause 65, the Project Manager may instruct the Contractor to submit a quotation for a proposed instruction. Note that the Contractor has only been instructed to submit a quotation; he does not carry out the work until instructed to do so.

The Contractor submits his quotation within three weeks of being instructed to do so.

Once the Contractor has submitted his quotation to the Project Manager, the Project Manager is required to reply as follows:

- he could instruct the Contractor to submit a revised quotation, in which case he explains his reasons for doing so to the Contractor. The Contractor then has a further three weeks to re-submit it, though in most cases he will review and re-submit within a shorter time scale;
- he could issue an instruction to carry out the work and accept the quotation, in which case it is implemented;
- he could decide not to proceed with a proposed instruction;
- he could notify the Contractor that the quotation is not accepted.

If the quotation is not accepted the Project Manager may still instruct the work to be carried out, but it is treated as a compensation event and the Contractor is instructed to submit a quotation.

6.14 Implementing compensation events

Under Clause 66, the Project Manager implements each compensation event by notifying the Contractor of the quotation which he has accepted or of his own assessment.

Where the Project Manager has not responded to the Contractor's notification on time (Clauses 64.4) then the event is implemented when the Contractor's quotation is deemed to be accepted.

This clause emphasises the finality of the assessment of compensation events.

If the subsequent records of resources on work actually carried out show that achieved Defined Cost and timing are different from the forecasts included in the accepted quotation or in the Project Manager's assessment, the assessment is not changed. The only circumstances in which a review is possible are those stated in Clause 61.6 (Project Manager's Assumption).

Value engineering

The contract provides for value engineering and sharing of savings between the Client and the Contractor under the following clauses.

Options A and B

Clause 63.12 – If the effect of a compensation event is to reduce the total Defined Cost and the event is a change to the Scope provided by the Client, which the Contractor proposed and the Project Manager accepted, the Prices are reduced by an amount calculated by multiplying the assessed effect by the value engineering percentage.

The value engineering percentage is stated in contract Data Part 1, the default being 50% unless a different percentage is stated.

Options C and D

Clause 63.13 – If the effect of a compensation event is to reduce the total Defined Cost and the event is a change to the Scope provided by the Client, which the Contractor proposed and the Project Manager accepted, the Prices are not reduced, i.e. the target is not reduced, so the Contractor gains the financial benefit through the Contractor's share.

Example

The Contractor proposed the use of precast concrete beams for a part of the works, which could give savings of £40,000. The Project Manager accepted the proposal and the Scope was changed to provide for the proposal.

If it were an Option A or B Contract

In this case, if the default 50% is in the Contract Data, the Prices are reduced by £40,000 × 50% = £20,000, effectively both parties benefiting equally by the saving.

If it were an Option C or D Contract

In this case, the Prices (the Target) is not reduced, but the Defined Cost plus Fee will be lower, so the Client and the Contractor effectively share the savings when the Contractor's share is calculated at Completion.

6.15 Final Certificate

The ECC does not have the equivalent of a Final Account or Final Certificate which is found in other contracts, certifying that the contract has fully and finally been complied with and that issues such as, in this case, compensation events have all been dealt with.

Some may say that because of the compensation event provisions there is no Final Account, as the implementation of the last compensation event closes the process, so there is no need for a Final Certificate, but some questions could still remain. See Chapter 5 (Final Assessment) for changes within the NEC4 ECC in this respect.

7   Title

7.1 Introduction

Title is covered within the ECC by Clauses 70.1 to 74.1.

The general principle with title to Plant and Materials brought onto a construction site by the Contractor is that, in the absence of any contractual provision to the contrary, title will remain with the Contractor until the Plant and Materials are incorporated (built) into the works, at which point they will become the property of the owner of the land upon which the works are being built. Once title has passed in this way the Contractor cannot, without an express right in the contract, normally the consent of the Contract Administrator, remove them.

7.2 Title provisions within the ECC

Within the ECC, title to Plant and Materials outside the Working Areas (off site) is covered by Clauses 70.1 and 71.1.

Clause 70.1 states that "whatever title the Contractor has to Plant and Materials which are outside the Working Areas passes to the Client if the Supervisor has marked them as for the contract". The term "marking" refers to the Supervisor either physically marking the Plant and Materials with some form of label or sign, or simply recognising that it exists by taking photographs and/or making an inventory of the Plant and Materials. This also ties in with Clause 71.1 regarding payment for Plant and Materials outside the Working Areas.

Note that in saying that "whatever title" the Contractor has, if it does not have title itself, normally because it has not paid for it, then it cannot pass title to the Client.

Title to Plant and Materials within the Working Areas (on site) is covered by Clause 70.2 which states that "whatever title the Contractor has to Plant and Materials which is outside the Working Areas passes to the Client if they have been brought within the Working Areas". Again, in using the words "whatever title the Contractor has", it cannot pass title which it does not hold itself.

Under Clause 70.2, if the Contractor removes Plant and Materials from the Working Areas with the Project Manager's permission title reverts back to the Contractor.

Two queries immediately come to mind with this particular provision:

(i) One could take the requirement for permission to its extreme in that the Project Manager has to give permission for the Contractor to remove each item of redundant or waste material from the Site!

Certainly, the definition of Plant and Materials under Clause 11.2(14) refers to "intended to be included in the works", but waste materials for example, were originally intended, but subsequently were never used within the works! It is the intention, not the outcome that is stated.

(ii) A Subcontractor or supplier would not be a party to the contract, so it would not need the Project Manager's permission to remove materials, however if it was appointed on the Engineering and Construction Subcontract it would need the Contractor's permission.

Title will then pass to the Client once the Plant and Materials are incorporated into the works.

Note that the ECC has no express right for the Contractor to be paid for unfixed Materials either within or outside the Working Areas (normally referred to in other contracts as "unfixed materials on site" or "materials off site"). One has to refer to the specific Main Option clauses to determine any entitlement, for example under Options C to F, payment to the Contractor is based on Defined Cost paid by the Contractor before the next assessment date plus the Fee. In respect of the Defined Cost of Plant and Materials, where those Plant and Materials are being stored is not relevant to payment entitlement.

When writing the Scope and/or Z clauses when making provision for payment for unfixed Plant and Materials off site and in order to protect the Client's interests, one should also consider including the following criteria, probably as part of the "marking" process:

The Plant and Materials:

- are intended to be incorporated into the works.
- are as specified within the contract.
- are in their final state in that nothing remains to be done to them prior to them being incorporated into the works, for example a steel staircase should be fabricated and ready for delivery so it looks like a steel staircase, not a pile of unidentifiable stock steel.
- have been set aside from other stock, and labelled as destined for the specific project to which the contract relates.
- are vested in the Client, normally activated when the Client makes payment, which then transfers title in accordance with the contract, free from any charges or encumbrances.

In reality, vesting certificates are only effective during the transition period whilst everyone in the supply chain awaits payment. If someone in the supply chain becomes insolvent and those below them do not get paid, then title belongs to the "last man in the chain" who has not been paid (see also "retention of title" below).

- are insured against loss or damage for their full value under a policy of insurance protecting the interests of the Client and the Contractor until they are delivered to the Site (the Working Areas).
- can be inspected at any time.
- cannot be removed from the storage location, except for delivery to Site (the Working Areas).

The issue of title to Plant and Materials does not tend to cause many problems unless a party becomes insolvent; this then becomes a major problem as a party such as a Client who believed it had good title to unfixed Plant and Materials because the contract stated that when they were on Site it had title, finds that it does not have title.

Whilst this book is intended for international use rather than purely focusing on English law and case law, it is not unknown for a Client to pay a Contractor for unfixed materials on site, but due to the insolvency of a Subcontractor and the Contractor not having title to those materials, the Client then also had to pay the Subcontractor!

7.3 Retention of title

One aspect to bear in mind in considering the words "whatever title the Contractor has to Plant and Materials" is the use of "retention of title" clauses.

A retention of title clause, which normally states "title of the goods remains with the seller until the purchase price is paid in full" is one that allows the supplier of materials to retain ownership of those materials until they are fully and finally paid for, irrespective of any contractual provisions or agreements between say, the Contractor and the Client. Sometimes a retention of title clause may state "title in the goods remains with the seller until the price and all other sums owing by the purchaser to the seller are paid in full".

Such clauses will override any clauses in the main contract that may stipulate that title will pass to the Client when materials are delivered to the site, when they are marked, or when the Client has paid for them as the supplier is not party to the contract. These clauses then provide security for the supplier against the purchaser's insolvency.

In order to be effective, retention of title clauses must be stipulated and accepted at the time the contract is formed, i.e. when the supplier's quotation is offered and accepted.

In order to provide this protection when insolvency occurs, the receiver or liquidator must be satisfied on each of the following:

- that the wording of the clause includes the specific goods for which protection is being claimed,
- that the clause has been incorporated into the contract for the supply of those materials,
- that any goods claimed can conclusively be identified, and
- that the goods supplied relate to an unpaid invoice.

The goods being claimed under the retention of title clause must be identifiable and the receiver must be fully satisfied that any items claimed are the actual goods supplied by the supplier claiming them. Most retention of title clauses relate to raw materials, stock in trade or livestock.

The best method of identification is where the goods are marked with the name of the supplier or where their serial numbers are quoted on any unpaid invoices. The supplier should be allowed access to the insolvent's premises to inspect the goods which it considers are subject to its claim. Of course, the supplier should not be allowed to remove any goods until the official receiver is satisfied that the claim is valid.

If a retention of title clause was drafted only to retain ownership of the goods until payment was made for them, goods such as any raw materials cease to be caught by it once the manufacturing process has begun, i.e. once the goods have lost their identity. For example, leather supplied in the production of handbags, once cut is deemed to have created a new product.

It may be possible for the supplier to retain title to the goods supplied even if they have been used in a manufacturing process provided they are still identifiable, in their original form and are easily removable. For instance, ceiling tiles supplied to be installed into a suspended ceiling do not lose their identity because they are fixed within the ceiling grid. However they must be capable of being removed without damaging the ceiling grid.

7.4 Objects of value

In the event that the Contractor discovers an object of value or of historical or other interest, such as archaeological remains within the Site, it has no title to it. It notifies the Project Manager, who then instructs the Contractor how to deal with it. It is vital that the Contractor only takes instructions form the Project Manager in this respect, although in the case of archaeological remains, other parties such as local authorities, museums and educational establishments may have an interest in the find and how it is dealt with.

This is a compensation event under Clause 60.1(7) so it is important that the Contractor keeps detailed records of any stoppages, disruption and any changes to working methods as a result of the find.

Unless otherwise stated in the Scope, the Contractor has the title to materials from excavation or demolition (Clause 73.2). This is more practical than the NEC3 ECC which was the reverse, in that the Contractor only had title as stated in the Works Information (now "Scope").

It is important in this respect that the Scope clearly details any title the Client has to materials as it may wish to sell, recycle or remove those materials, and also if the Client has title to the materials, it must clearly state what the Contractor is to do with those materials, for example hand them to the Client on site, transport them to another location, etc.

Finally, the Contractor has the right to use material provided by the Client only to Provide the Works, the right also being available to a Subcontractor.

8 Liabilities and insurance

8.1 Introduction

Liabilities and insurance are covered within the ECC by Clauses 80.1 to 86.3.

Insurance is the principle of many paying in for the few to be compensated. It is basically a contract between the "insured" who offers to pay an agreed sum, a premium, to the "insurer" who warrants to pay out an agreed sum should a particular and specified event occur, e.g. damage to property, death or injury, etc. In effect, it is possible to insure against anything which holds an element of risk.

The required insurances under a construction contract can be grouped under the following headings:

(i) Loss or damage to the works, Plant, and Materials

This covers loss or damage to the works in progress, any unfixed Plant and Materials, and any Equipment on site.

(ii) Public Liability

This is to cover the insured against any death, injury or damage claims from parties other than their own employees.

(iii) Employer's Liability

In many countries, employers of staff have a statutory obligation to provide cover for their employees in the event of death or injury caused during the course of their employment. "Worker's Compensation" can also form part of this insurance.

8.2 Liabilities and insurance provisions within the ECC

Clause 80.1 lists the Client's liabilities, however, these are the Client's liabilities only in terms of loss, wear or damage, and are not an exhaustive list of all the liabilities the Clients bears under the contract. There are many other risks such as actions or inactions of the Project Manager or Supervisor, unforeseen physical conditions and weather, which are covered elsewhere as compensation events.

There are nine main categories of Client's liability within this clause:

(i)   Claims and proceedings relating to the Client's use or occupation of the Site, negligence, breach of statutory duty or interference with any legal right, or a fault in its design.

(ii)   A fault of the Client or any person employed by it.

(iii)   A fault in the design contained in the Scope provided by the Client or an instruction from the Project Manager changing the Scope.

In respect of design, the Client should cover the risk itself through professional indemnity insurance, or if it uses external consultants for the design, it should ensure that they hold such insurance. This insurance should be held for the full period of liability, which will normally be several years post completion of the works.

(iv)   Loss or damage to Plant or Materials supplied by the Client to the Contractor or by Others on the Client's behalf until the Contractor has received and accepted them.

(v)   Loss or damage to the works, Plant and Materials, caused by matters outside the control of the Parties.

(vi)   Loss or damage to parts of the works taken over by the Client, except where due to a Defect that existed at takeover, an event occurring before takeover which was not a Client's liability, or due to the activities of the Contractor on the Site after takeover.

(vii)   Loss or damage to the works and any Equipment, Plant and Materials retained on Site by the Client after termination, except loss or damage due to the activities of the Contractor on the Site after termination.

(viii)   Loss or damage to property owned or occupied by the Client other than the works, unless arising from or in connection with the Contractor Providing the Works.

(ix)   Additional Client's liabilities stated in the Contract Data.

Clause 81.1 lists the Contractor's liabilities:

(i)   Claims and proceedings from Others, and compensation and costs payable to Others which arises from the Contractor Providing the Works.

(ii)   Loss of or damage to the works, Plant and Materials and Equipment.

(iii)   Loss of or damage to property owned or occupied by the Client other than the works which arises from the Contractor Providing the Works.

(iv)   Death or bodily injury to employees of the Contractor.

This is a new provision, previous NEC contracts stating that any risk (now "liability") not carried by the Employer (now "Client") is carried by the Contractor.

8.3 Recovery of costs

- any cost which the Client pays as a result of an event for which the Contractor is liable is paid by the Contractor.
- any cost which the Contractor pays as a result of an event for which the Client is liable is paid by the Client.

The right to recovery of such costs is reduced if the recovering Party is partly liable for the costs.

8.4 The Insurance Table

The Insurance Table in the contract itemises the insurances that the Contractor has to effect, together with the minimum amount of cover.

The default position is that the Contractor provides, maintains and pays for the insurances, unless the Client states otherwise in Contract Data Part 1.

Whether the Contractor or the Client takes out the insurances, they are always effected as a joint names policy, except the fourth insurance, which is Employer's Liability Insurance. The insurances are to be effective from the starting date until the Defects Certificate or a termination certificate has been issued.

The four insurances listed within the Insurance Table are:

(i) Loss or damage to the works, Plant and Materials

Insurance of the Works will normally be covered by the Contractor's All Risks (CAR) policy. The minimum amount of cover for all insurances is stated in the Insurance Table, however the Contractor is liable for whatever the amount of any claim therefore it must consider the minimum value in the Insurance Table purely as a guide.

(ii) Loss of or damage to Equipment

Again, the Contractor's All Risks (CAR) policy should cover this. The reference to "replacement cost" means the cost of replacement with Equipment of similar age and condition rather than "new for old".

(iii) Loss of or damage to property (except the works, Plant or Materials) or bodily injury or death not of an employee of the Contractor arising in connection with the Contractor Providing the Works.

This requires the Contractor to indemnify the Client against any loss, expense, claim, etc in respect of any personal injury or death caused by the carrying out of the work, other than their own employees. This includes the liability toward members of the public who may be affected by the construction work, although they have no part in it. In the case where a party makes a claim directly against the Client due to a death or injury the Contractor should either take on that claim, or alternatively the Client can sue the Contractor to recover any monies.

The Insurance Table states the minimum amount of cover or minimum limit of indemnity. The Contractor could be liable for whatever the amount of any Claim therefore it must consider the minimum value in the Insurance Table purely as a guide.

(iv) Death of, or bodily injury to, employees of the Contractor

As stated above, this is the only insurance within the Insurance Table that is not effected in the joint names of the Contractor and the Client. In many countries this insurance is a legal requirement for a company that employs people.

This covers the Contractor's liabilities as an employer of people to insure against injury or death caused to those employees whilst carrying out their work, which is a legal obligation in most countries.

Again, the Insurance Table states the minimum amount of cover or minimum limit of indemnity, the Contractor having to insure for whatever the amount of any claim.

8.5 What is Contractors' All Risks insurance?

Contractor's All Risks insurance normally covers damage to property, such as damage to buildings, infrastructure or other structures being worked on. It also covers liability for third party claims for injury and death or damage to third party property.

This insurance is usually taken out in the joint names of the Contractor and the Client. Other interested parties, such as funders, often ask to be added as a joint name. The theory is that if damage occurs to the insured property then, regardless of fault, insurance funds will be available to allow for reinstatement.

The effect of Joint Names Insurance is that each party has its own rights under the policy and can therefore claim against the insurer. Each insured should comply with the duties of disclosure and notification.

8.6 Insurance of neighbouring property

Some contracts have provision for insuring neighbouring property which may suffer from collapse, subsidence, vibration, etc. due to damage by, for example, piling, deep excavations or demolition works, where not caused by the negligence of the Contractor.

The ECC does not provide for this, but it can be added as a Z clause if applicable.

The Client should consider its liabilities in respect of neighbouring owners who may be affected by the works and if necessary either instruct the Contractor to provide such insurance or take it out itself.

Such a policy would normally exclude:

(i) Damage caused by the negligence of the Contractor or its Subcontractors.
(ii) Damage which can be reasonably foreseen, e.g. cracking may have already occurred to a neighbouring building prior to the works commencing. In this

respect, it is important that a condition survey with photographic evidence be carried out before construction commences.

(iii) Where the damage is caused by any excepted risks identified in the contract. These risks will normally be held by the Client.

8.7 Professional Indemnity (PI) Insurance

Contract Data Part 1 provides for the Client to insert a requirement for the Contractor to provide additional insurance.

Where the Contractor is designing parts of the works, it would be advisable for the Client to include a requirement for Professional Indemnity (PI) Insurance.

Particularly if a Contractor is providing design and makes a mistake, is found to be negligent, or gives inaccurate advice, then it will be liable to the Client in event that the Client incurs a loss as a result. This loss can be very significant where the design has to be corrected, parts of the structure have to be taken down and reinstated, a facility has to be closed down whilst the remedial measures take place, and there are also legal costs.

Professional Indemnity claims can arise where there is negligence, misrepresentation or inaccurate advice which does not give rise to bodily injury, property damage or personal injury, but does give rise to some financial loss.

Additional coverage for breach of warranty, intellectual property, personal injury, security and cost of contract can be added.

In that event, although the Client claims against the Contractor or Consultant rather than from the insurers, Professional Indemnity Insurance protects the Contractor or Consultant against claims for loss or damage made by a client or third party.

Professional Indemnity Insurance policies are based on a "claims made" basis, meaning that the policy only covers claims made during the policy period when the policy is "live", so claims which may relate to events occurring before the coverage was active may not be covered. However, these policies may have a retroactive date which can operate to provide cover for claims made during the policy period but which relate to an incident after the retroactive date. The Project Manager, on behalf of the Client should ensure that the Contractor has taken out and maintained the required insurances for the full period of its liability.

Claims which may relate to incidents occurring before the policy was active may not be covered, although a policy may sometimes have a retroactive date, earlier than when the policy was created.

8.8 Proof of insurance

The Contractor is required to submit to the Project Manager for acceptance certificates of insurance as required by the contract signed by the insurer or the insurance broker:

(i) Before the starting date
(ii) On each renewal of the insurance policy

If the Contractor does not submit a required certificate, the Client may insure a risk, the cost of the premium being paid by the Contractor.

If the Client is to provide insurance, then the Project Manager submits certificates to the Contractor for acceptance before the starting date and as and when instructed by the Contractor.

If the Client does not submit a required certificate, the Contractor may insure a risk, the cost of the premium being paid by the Client.

9 Termination

9.1 Introduction

Termination is covered within the ECC by Clauses 90.1 to 93.2.

Termination is a word most commonly used in the context of construction contracts to refer to the ending of the Contractor's employment. The parties have a common law right to bring the contract to an end in certain circumstances, but most standard forms give the parties additional and express rights to terminate upon the happening of specified events.

Some contracts refer to termination of the contract, whilst others refer to the termination of the Contractor's employment under the contract. In practice, it makes little difference, because most contracts make express provision for what is to happen after termination.

The author generally advises against termination, unless there is no other option for the Parties, for example insolvency of a Party, where there is no choice. This is because the administration of the termination process can be very complex.

The Parties should therefore always view termination as an entitlement not an obligation.

It is also important to note that neither party should attempt to terminate employment under the contract unless they are sure that the provision is available within the contract for a particular default and they follow the provisions "to the letter".

If termination is held to be wrongfully applied, it is usually considered a repudiation of the contract.

9.2 Termination provisions within the ECC

Within the ECC, there are 22 reasons listed for either party to terminate:

- **Reasons R1 to R10** provide for either Party to terminate due to the insolvency of the other. This includes bankruptcy, appointment of receivers, winding-up orders and administration orders dependent on whether the party is an individual, a company or a partnership.

- **Reasons R11 to R13** provide for the Client to terminate if the Project Manager has notified the Contractor that it has defaulted and the Contractor has not remedied the default within four weeks of the notification of default in respect of the Contractor substantially failing to comply with its obligations, not providing a bond or guarantee, or appointing a Subcontractor for substantial work before the Project Manager has accepted the Subcontractor.

    It is quite subjective as to what the words "substantially failed to comply" and "substantial work" mean within these provisions. One would assume that it is not intended to relate to minor breaches, but the Contractor could "substantially fail to comply" with a minor obligation.

- **Reasons R14 to R15** provide for the Client to terminate if the Project Manager has notified that the Contractor has defaulted and the Contractor has not stopped defaulting within four weeks of the notification in respect of substantially hindering the Client or Others, or has substantially broken a health or safety regulation.

    Again, the use of the word "substantially" is quite subjective. How would one substantially hinder someone, or substantially break a health or safety regulation? Isn't a health or safety regulation either "broken" or "not broken"?

- **Reason 16** provides for the Contractor to terminate if the Client has not paid an amount due under the contract, within thirteen weeks of when it should have been paid.

- **Reason 17** provides for either Party to terminate if the Parties have been released under the law from further performance of the whole of the contract. Note the reference to "the whole of the contract", not just a part.

- **Reasons 18 to 20** provide for the Project Manager having instructed the Contractor to stop or not restart any substantial work or all work and an instruction allowing the work to restart or start has not been given within thirteen weeks.

    Either Party may terminate if the instruction was due to a default by the other, or if the instruction was due to any other reason.

- **Reason R21** provides for the Client to terminate if an event occurs which stops the Contractor completing the works, or stops the Contractor completing the works by the date shown on the Accepted Programme and is forecast to delay Completion by more than thirteen weeks, and which neither Party could prevent and an experienced Contractor would have judged at the Contract Date to have such a small chance of occurring that it would have been unreasonable for him to have allowed for it.

- **Reason R22** provides for the Client to terminate if the Contractor does a Corrupt Act, unless it was done by a Subcontractor or supplier, and the Contractor was not and should not have been aware of the Corrupt Act, or the Contractor informed the Project Manager of the Corrupt Act and took action to stop it as soon as it became aware of it.

Note that a Corrupt Act is defined under Clause 11.2(5) as "*offering, promising, giving, accepting or soliciting of advantage as an inducement for an action which is illegal, unethical or a breach of trust or abusing any entrusted power for private gain . . . in connection with this contract, or any other contract with the Client.*"

With all the termination reasons, the Party wishing to terminate notifies the Project Manager and the other Party giving details of its reasons for terminating. The Project Manager then issues a termination certificate to both parties if it is satisfied that the reason for termination is valid under the contract. Is it perhaps curious that the Project Manager, who acts for the Client makes the decision as to whether the termination is valid?

Once the termination certificate is issued the termination process commences and the Contractor is not required to carry out anything further in Providing the Works.

Note that if the Client terminates under Reasons R1 to R15, R18 or R22, if a certified payment has not been made at the date of the termination certificate, the Client need not make the payment unless the contract states otherwise. This is because there is likely to be a payment due from the Contractor to the Client, so payments are "frozen".

Procedures on Termination

Under any of the reasons, whoever terminates, and whether the reason is included in the Termination Table or not, the Client may complete the works, either itself or using another Contractor, and may use any Plant and Materials to which it has title (Procedure P1).

Client Terminates

- for reasons R1 to R15, R18 or R22 and for a reason other than that stated in the Termination Table, the Client may instruct the Contractor to leave the site and remove any Equipment, Plant or Materials which belong to the Contractor and assign the benefit of any subcontract to the Client (Procedure P2).
- for reasons R1 to R15 or R17, R18, R20 or R22, the Client may use any Equipment to which the Contractor has title and complete the service. The Contractor promptly removes such Equipment from Site when informed to do so.

    Clearly in this respect, the Contractor does not have title to hired Equipment; also if the Equipment is owned by the Contractor but it has become insolvent, then some discussion may have to take place between the Client and the relevant insolvency practitioner or other authorised representatives (Procedure P3).
- for reason R21, the Contractor leaves the Working Areas and removes the Equipment (Procedure P4).

Contractor Terminates

- whatever the reason, the Contractor follows procedures P1 and P4 which are described above.

Payment on Termination

If the Client or the Contractor terminates for any reason, the Contractor is entitled to be paid:

Amount A1

- an amount due assessed as for normal payments;
- the Defined Cost for Plant and Materials within the Working Areas to which the Client has title and of which the Contractor has to accept delivery;
- other Defined Cost in reasonable expectation of completing the whole of the works;
- any amounts retained by the Client; and
- deduction of any un-repaid balance of an advanced payment.

If the Client terminates for reasons R17, R20 or R21, or the Contractor terminates for any reason, the Contractor is entitled to be paid:

Amount A2

- the forecast Defined Cost of removing Equipment.

If the Client terminates for reasons R1 to R15, R18 or R22, the Client is entitled to be paid:

Amount A3

- a deduction of the forecast of the additional cost to the Client of completing the whole of the works.

If the Contractor terminates for reasons R1 to R10, R16 or R19, the Contractor is entitled to be paid:

Amount A4

The fee percentage applied to

- for Options A, B, C or D – any excess of the total of the Prices at the Contract Date and the Price for Work Done to Date.
- for Options E and F – any excess of the first forecast of the Defined Cost over the Price for Work Done to Date less the Fee.

Under Options C and D, the Project Manager assesses the Contractor's share after certifying termination.

The Price for Work Done to Date is based on:

- Defined Cost which the Contractor has paid, and is committed to pay for work done prior to the termination.

and also, under Option C:

- the total of the Prices based on the lump sum price for each completed activity, and a proportion of the lump sum for each incomplete activity.

The Contractor's share is added to the amount due to the Contractor on termination.

9.3 Termination and Subcontractors

The Parties to the ECC should also consider how termination of their contract affects Subcontractors. If the Contractor is the defaulter, termination occurs and the Client wishes to proceed with the works, any subcontracts would be assigned to the Client.

However, if the Contractor is the terminating party it should ensure that provisions are in place regarding termination of subcontracts, and termination provisions in the main contract should be mirrored in the subcontracts.

10 Resolving and avoiding disputes

10.1 Introduction

Resolving and avoiding disputes is covered within the ECC by Options W1, W2 and W3.

Options W1 and W2 cover the adjudication process, whether the UK-based legislation, the Housing Grants, Construction and Regeneration Act 1996 is applicable (Option W2), or not applicable (Option W1). The Act applies in England, Wales and Scotland. Similar legislation of a different title applies in Northern Ireland.

Note that the Housing Grants, Construction and Regeneration Act 1996 is supplemented by the Local Democracy, Economic Development and Construction Act 2009.

If the ECC is used within the UK, the Client must by reference to the Act determine whether the contract is a "construction contract", and select the appropriate dispute resolution Option. If the ECC is used outside the UK, Option W1 will always apply.

Clearly, there is nothing to prevent the parties jointly attempting to resolve a dispute as they see fit, by a consensual and non-binding route such as negotiation, mediation, conciliation or mini trial, prior to referring it to adjudication.

10.2 What is adjudication?

Adjudication has been defined as *"a summary non-judicial dispute resolution procedure that leads to a decision by an independent person that is temporarily binding but which may subsequently be reviewed by means of arbitration, litigation or by agreement"*.

It is therefore an intermediate dispute resolution process which is binding upon the parties unless and until the matter is referred to a tribunal, normally arbitration or litigation.

In the UK, adjudication is a statutory right, but it has always been included as the primary means to resolve a dispute within the NEC contracts, failing which the parties may refer a dispute to a tribunal. In fact, the majority of adjudication decisions are accepted by parties as the final result and are never taken to the tribunal.

10.3 Who can be an Adjudicator?

Whilst many professional bodies hold registers of Adjudicators there is no formal qualification to become an Adjudicator, though there are commonly acceptable requirements which include:

- a recognised professional qualification in a relevant discipline. The relevant discipline then relates to the dispute in question, for example if the dispute is related to services installations, a Mechanical or Electrical Engineer would probably be a relevant discipline, though it must be remembered that the Adjudicator's role is to enforce the contract, so in that case it does not necessarily have to be a Mechanical or Electrical Engineer just because the dispute is related to services. Clearly, as disputes are always of a legal nature, lawyers tend to make up a good proportion of the adjudicating community.
- substantial number of years' experience within the industry, normally 10 to 15 years post qualification in a relevant discipline.
- a qualification, or at least a thorough understanding of, contracts and legal principles, though in many cases the Adjudicator is not a lawyer, nor does it have to be. Many say it is better if it is not a lawyer, but that is probably a view left to the cynics!
- recognised training in the law and practice of adjudication.
- membership of a professional institution.
- some also say that the Adjudicator should have Professional Indemnity (PI) insurance, though it must be emphasised that as the Adjudicator is not liable to the Parties for any action or inaction on its part other than in bad faith, it could be questioned as to whether such insurance should be a prerequisite to appointment. Most adjudicator nominating bodies require their panel members to carry appropriate insurance.

As the use of adjudication in the resolution of contractual disputes has grown, particularly in the UK where parties have a statutory right to refer a dispute to adjudication, many education establishments and professional bodies now provide diploma and certificate courses in adjudication.

10.4 Who pays for the adjudication?

The cost of the Adjudicator's fee is allocated by him/her between the parties, the Adjudicator deciding what proportion is to be paid by each party, with in some cases one party paying the majority or even all of the Adjudicator's fee. However, both parties are jointly and severally liable for payment of the Adjudicator's fee, therefore if one party will not or cannot pay its share, possibly due to insolvency, the other party must pay it, and recover the amount directly from the defaulting party.

Each party bears its own costs in professional fees, collating, copying, postage, etc which is different to, for example, litigation where both parties' costs are borne by the losing party.

10.5 What will the Adjudicator charge?

There are no set hourly rates for Adjudicators; each Adjudicator will advise the parties of its hourly rate and whether there are other costs payable, for example for travel or printing. The Adjudicator is required to properly record the time spent carrying out its duties.

10.6 What if a party does not comply with the Adjudicator's decision?

If a party does not comply with the Adjudicator's decision, the other party can enforce that decision by going to court.

10.7 Selecting the Adjudicator

Within Contract Data Part 1 there are three options for selecting the Adjudicator:

(i) The name and address of the Adjudicator may be stated by the Client in Contract Data Part 1 and it is then appointed before the starting date.

This has the advantage that tendering Contractors are aware at the time of tender who will be the Adjudicator in the event that a dispute arises and, if necessary, can address the issue and object to them within their tender submissions. Also, the Adjudicator is already in place should a dispute arise. The named Adjudicator may require some form of retainer fee for being named in the contract and being available in the event that a dispute arises.

(ii) The parties can mutually agree to the name of the Adjudicator in the event that a dispute arises. This has the advantage that there is no "Adjudicator in waiting" and therefore no fee payable. Many ECC practitioners say that pre-appointing the Adjudicator as in option (i) is resigning oneself to the fact that there will at some point be a dispute. However, once parties are in dispute they then have to agree who is to be the Adjudicator and appoint him, though, at this point they may not wish to agree with each other about anything!

(iii) The name of the Adjudicator Nominating Body may be stated by the Client in Contract Data Part 1. This is probably the favoured option as an independent name can be put forward by the nominating body, who is normally well prepared to offer the name of an Adjudicator from its registers and have him appointed within the timescales set by the contract and if necessary the appropriate legislation.

In all cases, the Adjudicator must be impartial, i.e. the Adjudicator should not show any bias towards either party. In addition, all correspondence from the Adjudicator must be circulated to both parties.

Any request for an Adjudicator must be accompanied by a copy of the Notice of Adjudication, and the appointment of the Adjudicator should take place within seven days of the submission of the Notice of Adjudication to the other party.

10.8 Adjudicator nominating bodies

Adjudicator Nominating Bodies, as the name suggests, are organisations that fulfil the role of nominating Adjudicators. These bodies keep registers of Adjudicators with varying expertise and based at various geographical locations who can act for parties should they be nominated.

Clearly, in naming the Adjudicator Nominating Body in the contract it is advisable to name an organisation with expertise in the project to be carried out.

These bodies are very knowledgeable about appointment of Adjudicators and the relevant timescales and for a modest fee can nominate a suitably qualified Adjudicator to suit the parties' requirements.

10.9 Preparing for adjudication

Once the adjudication process starts by the issuing of a Notice of Adjudication by the referring party it is a very quick procedure, therefore it is critical that the referring party, and in turn the responding party, prepare well in the short time they have available. Timescales are stated within the contract, though these may be extended by agreement between the parties.

As the process does not start until the referring party initiates the process by issuing the Notice of Adjudication to the other party, the referrer does have the advantage of time, and in many cases surprise, as it can prepare its case before issuing the notification.

Before considering initiating the adjudication process, parties should consider the following:

(i) There must be a dispute – whilst this may seem fairly obvious it is essential that a dispute actually exists between the parties in that one party has made a claim, which the other party denies (probably repeatedly), and that the parties have exhausted the means to resolve the dispute themselves, for example by the use of negotiation, mediation, conciliation etc., though they do not have to prove that they have tried other methods.

(ii) The dispute arises under or in connection with the contract – the Adjudicator is appointed to decide a dispute under the contract, therefore it is essential that the dispute arises on a matter covered by the contract. If the dispute is whether a contract exists in the first place, then an Adjudicator cannot decide that.

(iii) The role of the Adjudicator is to enforce the contract, not to decide who is the more deserving party. It is therefore essential that, before commencing adjudication, that the referring party ensures that it has properly complied with the contract, otherwise it may find itself in a weak position.

There have been adjudications in the past where a Contractor has acted in good faith but not strictly in accordance with the contract, accepting and willingly acting on verbal instructions in order not to damage what appeared to be a good relationship with the Client, then when a dispute arises the Adjudicator states that as the contract did not provide for verbal instructions, the Contractor was incorrect to act on them, and therefore the Contractor loses the adjudication.

(iv)   Whilst many people say that, as adjudication is a fairly simple process, it should not include lawyers, as the issues are industry-related rather than legal issues, it is important that the parties carry out their obligations within the short timescale they have, so if a party is uncertain as to what they have to do, and when and how they have to do it, they obtain professional advice.

(v)   It is essential that a clear and concise notice of adjudication is prepared by the referring party as it defines what is in dispute, and the matters that are being referred to the Adjudicator. The Adjudicator cannot consider issues which are not set out within the Notice of Adjudication.

The notice should include:

*   Brief details of the project.
    Whilst the Adjudicator may have been named in the contract, it may have little knowledge about the actual project.
*   The names and addresses of the parties to the contract and the parties in dispute.
*   The nature and a brief description of the dispute.
*   The nature of the redress which is sought from the Adjudicator.

It is critical that the responding party, and also the Adjudicator, clearly understand what redress the referring party is seeking by referring the matter to adjudication.

(i)   Details included within the referral should be clearly and concisely presented so that the responding party knows what it needs to respond to, and also the Adjudicator clearly understands what is required of him. Any supporting information should be clearly referenced and relative to the matter in dispute.

Neither party should include or rely on any information or evidence which the other party has not seen. If they do, the other party may raise an objection, which could bring the adjudication process to a premature end without a resolution.

(ii)   The contract provides for timescales to be extended by agreement between the parties, so if more time is needed in order to prepare and submit documentation, the other party should be consulted.

10.10 Statutory right to adjudication (UK-based contracts)

When the Housing Grants Construction and Regeneration Act 1996 was introduced in the UK, it applied to construction contracts which came into existence after 1 May 1998 and, apart from a few notable exceptions, it provided a statutory right to refer a dispute under a construction contract to adjudication. This Act has now been amended by the Local Democracy, Economic Development and Construction Act 2009.

The Act applies to all that it terms "construction contracts". That term captures many activities that feature in construction contracts even though construction in its normal definition might not be involved, including:

- all normal building and civil engineering work, including construction, alteration, repair, maintenance, extension and demolition or dismantling of structures forming part of the land and works forming part of the land, whether they are permanent or not;
- the installation of mechanical, electrical and heating works and maintenance of such works;
- cleaning carried out in the course of construction, alteration, repair, extension, painting and decorating and preparatory works;
- agreements with consultants such as architects, other designers, engineers and surveyors;
- elements such as scaffolding, site clearance, painting and decorating;
- labour-only contracts;
- contracts of any value.

It excludes:

- work on process plant and on its supporting or access steelwork on sites where the primary activity is nuclear processing, power generation, water or effluent treatment, handling of chemicals, pharmaceuticals, oil, gas;
- supply-only contracts, that is, contracts for the manufacture or delivery to site of goods where the contract does not provide for their installation;
- extracting natural gas, oil and minerals (and the workings for them);
- purely artistic work;
- off-site manufacture;
- contracts with residential occupiers – this means a contract where work is carried out on a dwelling which one of the parties occupies or intends to occupy as its residence rather than simply a housing project;
- PFI contracts – this only includes the head agreements in PFI projects, not a contract for the construction of the works;
- finance agreements – this includes loan agreements.

The Adjudicator is appointed jointly by both parties who are jointly and severally liable for those fees.

Under the Act, the contract must include an adjudication clause which complies with the requirements of the Act. If the contractual clause does not comply then it is voidable and the parties can exercise their right to adjudication under the Scheme for Construction Contracts to be issued by the Secretary of State.

The provisions in the contract must:

- enable the parties to give notice at any time;
- provide a timetable with the object of securing the appointment of the adjudicate and referral of the dispute to him within seven working days;
- require the Adjudicator to reach a decision within twenty-eight days of referral or such longer period as is agreed by the parties after the dispute has been referred;
- allow the Adjudicator to extend the period of twenty-eight days by up to fourteen days with the consent of the party by whom the dispute was referred;
- impose a duty on the Adjudicator to be impartial;
- enable the Adjudicator to take the initiative in ascertaining the facts and the law;
- provide that the decision of the Adjudicator is binding until determined by litigation, arbitration or by agreement;
- provide that the Adjudicator is not liable for anything done or omitted in the discharge or purported discharge of its functions as Adjudicator unless the act or omission is in bad faith, and that any employee or agent of the Adjudicator is similarly protected from liability.

10.11 Does a "dispute" exist?

Only a dispute between the contracting parties can be referred to adjudication. If there is no dispute, then an Adjudicator has no jurisdiction to consider the matter.

As an example, if the Contractor notifies a compensation event and the Project Manager who represents the Client does not agree with it, then the simple fact that the claim is not agreed does not mean that there is a dispute as the parties are still free to meet and, if necessary, to negotiate a settlement. It is only when the parties have exhausted all the opportunities to meet and reach a compromise, but still have a difference, that a dispute exists, and the matter can be referred to the Adjudicator.

It must also be noted that an Adjudicator only has jurisdiction to consider one dispute at a time under the same contract unless there is agreement between the disputing parties. Therefore, if there are number of issues in dispute, unless they are inextricably joined then there is likely to be several adjudications.

Disputes under or in connection with the contract may be referred to the Adjudicator.

Note that a dispute under the contract cannot be referred to the tribunal (arbitration or litigation) unless it has first been decided by the Adjudicator.

Adjudication is a quick, convenient and relatively inexpensive way of resolving a dispute, whereby an impartial third-party Adjudicator either named in the contract, selected by agreement between the parties or nominated by a third party, decides the issues in dispute between the parties. It is quicker and less expensive than arbitration or litigation, though the parties may refer to one of those methods following adjudication.

It must be recognised that the role of the Adjudicator is to enforce the contract, not to decide what, in its opinion, would be the fairest outcome. For example, if the parties have a dispute over the quality of the replacement of suspended ceiling tiles carried out, it is the Adjudicator's task to ascertain whether the tiles were supplied and installed in accordance with the contract, i.e. specifications, etc. and the applicable law, not whether it was a good or bad job. In addition, the question of whether a contract exists cannot be decided by the Adjudicator but more likely through the courts.

10.12 Resolving and avoiding disputes provisions within the ECC

As stated above, the ECC provides two Options for referral to the Adjudicator:

Option W1

Option W1 is used when the UK Housing Grants, Construction and Regeneration Act 1996 does not apply.

Any dispute arising under the contract is, in accordance with the Dispute Reference Table, first referred by a Party issuing a notice to the Senior Representatives and copied to the other Party and the Project Manager stating the nature of the dispute it wishes them to resolve.

Each Party then submits their statement of case (no more than ten sides of A4 paper) with supporting evidence.

The Senior Representatives then try to resolve the dispute within three weeks, using any procedure they wish to and attending as many meetings as required to resolve the dispute. At the end of the three weeks, a list is issued showing the issues agreed, and the issues not agreed. The Project Manager and the Contractor put the agreed issues into effect.

According to the Dispute Resolution Table, if the Dispute is about:

- an action or inaction of the Project Manager or the Supervisor, either Party may refer it to the Senior Representatives, not more than four weeks after the Party became aware of the action or inaction;
- a programme, compensation event, or quotation for a compensation event which is treated as having been accepted, the Client may refer it to the Senior Representatives, not more than four weeks after it was treated as accepted;
- an assessment of Defined Cost which is treated as correct, either Party may refer it to the Senior Representatives, not more than four weeks after the assessment was treated as correct;
- any other matter, either Party may refer it to the Senior Representatives, when the dispute arises.

Note that if the referring Party is the Client and the dispute is regarding a compensation event which has been treated as having been accepted (Clause 62.6) then the Client must notify the dispute to the Contractor and then refer it to the Senior Representatives not more than four weeks after it was treated as accepted.

This seems an odd reason for referring a dispute to the Senior Representatives, as Clause 62.6 states that if the Project Manager does not reply to the Contractor's notification, it is treated as acceptance of the quotation; how could the Senior Representatives (or an Adjudicator) whose role is to enforce the contract find other than in favour of the Contractor?

Surely the Client's dissatisfaction, and the dispute itself, is with its Project Manager, not with the Contractor?

Whilst these timescales are fixed within the Dispute Reference Table, they can be extended by the Project Manager if the Contractor and the Project Manager agree before either the notice or the referral is due. Note that if the matter in dispute is not notified and referred within these timescales, neither Party can refer the dispute to the Adjudicator or the tribunal.

The Adjudicator is appointed under the NEC4 Dispute Resolution Service Contract. The Adjudicator acts impartially, and if it resigns or is unable to act, the Parties jointly appoint a new Adjudicator. The referring Party obtains a copy and completes the Dispute Resolution Service Contract. If the Parties have not appointed an Adjudicator, either party may ask the nominating body to choose an Adjudicator.

No procedures have been specified for appointing a suitable person, and in practice a number of different methods have been used. Whatever method is used, it is important that both Parties have full confidence in its impartiality, and for that reason it is preferable that a joint appointment is made.

The Adjudicator should be a person with experience in the type of work included in the contract between the Parties and who occupies or has occupied a senior position dealing with disputes. It should be able to listen to and understand the viewpoint of both Parties.

Often the Parties delay selecting an Adjudicator until a dispute has arisen, although this frequently results in a disagreement over who should be the Adjudicator.

As noted, the selection of the Adjudicator is important, and it should be recognised that a failure to agree an Adjudicator means that a third party will make the selection without necessarily consulting the Parties.

The adjudication process commences by the referring Party issuing a notice of adjudication to the Project Manager and to the other Party within two weeks of the production of the list of agreed and not agreed issues.

The notice of adjudication gives a brief description of the dispute, details of where and when the dispute has arisen, and the nature of the redress which is sought.

The Party referring the dispute to the Adjudicator must include within its referral information which it wishes to be considered by the Adjudicator, with any more information from either Party to be provided within four weeks of the referral.

A dispute under a subcontract, which is also a dispute under the ECC, may be referred to the Adjudicator at the same time, and the Adjudicator can decide the two disputes together.

The Adjudicator may review and revise any action or inaction of the Project Manager, take the initiative in ascertaining the facts, and the law relating to the dispute, instruct a party to provide further information and instruct a party to take any other action which it considers necessary to reach its decision.

All communications between a Party and the Adjudicator must be communicated to the other Party at the same time.

The Adjudicator decides the dispute and notifies the Parties and the Project Manager within four weeks of the end of the period for receiving information.

This four-week period may be extended by joint agreement between the Parties.

Until this decision has been communicated, the Parties proceed as if the matter in dispute was not disputed.

The Adjudicator's decision is temporarily binding on the parties unless and until revised by a tribunal and is enforceable as a contractual obligation on the parties.

The Adjudicator's decision is final and binding if neither Party has notified the Adjudicator that they are dissatisfied with an Adjudicator's decision within the time stated in the contract and intends to refer the matter to the tribunal.

The Adjudicator may, within two weeks of giving its decision to the Parties, correct a clerical mistake or ambiguity.

Review by the tribunal

A dispute cannot be referred to the tribunal unless it has first been referred to the Adjudicator.

The tribunal may be named by the Client within Contract Data Part 1.

Whilst no alternatives are stated, it will normally be litigation or arbitration.

If arbitration is chosen, the Client must also state in Contract Data Part 1, the procedure, the place where the arbitration is to be held, and the person or organisation who will choose an Arbitrator if the Parties cannot agree a choice, or if the named procedure does not state who selects the Arbitrator.

A Party can, following the adjudication, notify the other party within four weeks of the Adjudicator's decision that it is dissatisfied. This is a time-barred right, as the dissatisfied Party cannot refer the dispute to the tribunal unless it is notified within four weeks of the Adjudicator's decision; failure to do so will make the Adjudicator's decision final and binding.

Also, if the Adjudicator has not notified its decision within the time provided by the contract, a party may within four weeks of when the Adjudicator should have given its decisions notify the other party that it intends to refer the dispute to the tribunal.

The tribunal settles the dispute and has the power to reconsider any decision of the Adjudicator and review and revise any action or inaction of the Project Manager.

It is important to note that the tribunal is not a direct appeal against the Adjudicator's decision; the parties have the opportunity to present further information or evidence that was not originally presented to the Adjudicator, and also the Adjudicator cannot be called as a witness.

Option W2

Option W2 is used when the UK Housing Grants, Construction and Regeneration Act 1996 applies.

The inclusion of adjudication within the then New Engineering Contract, now the NEC4, predates UK legislation giving parties to a contract the statutory right to refer a dispute to adjudication.

The Housing Grants, Construction and Regeneration Act 1996

The Local Democracy, Economic Development and Construction Act 2009, which came into force in England and Wales on 1 October 2011 and in Scotland on 1 November 2011, amended the Housing Grants, Construction and Regeneration Act 1996, the new Act applying to most UK contracts after that date.

There still remains certain categories of construction contract to which neither Act applies.

The differences between the Housing Grants, Construction and Regeneration Act 1996 and the Local Democracy, Economic Development and Construction Act 2009 have been explored in a previous chapter but in respect of adjudication they include the following:

- The previous Act only applied to contracts which were in writing; the new Act also applies to oral contracts.
- Terms in contracts such as "the fees and expenses of the Adjudicator as well as the reasonable expenses of the other party shall be the responsibility of the party making the reference to the Adjudicator" will be prohibited.

  Much has been written about the effectiveness of such clauses, often referred to as Tolent clauses (*Bridgeway Construction Ltd vs Tolent Construction Ltd* 2000) and whether they comply with the previous Act, so the new Act should provide clarity for the future.

- The Adjudicator is permitted to correct its decision so as to remove a clerical or typographical error arising by accident or omission. Previously it could not make this correction.
- Section 108 of the Act provides parties to construction contracts with a right to refer disputes arising under the contract to adjudication. It sets out certain

minimum procedural requirements which enable either party to a dispute to refer the matter to an independent party who is then required to make a decision within twenty-eight days of the matter being referred.

- If a construction contract does not comply with the requirements of the Act, or if the contract does not include an adjudication procedure, a statutory default scheme, called the Scheme for Construction Contracts (referred to as the "Scheme"), will apply.

The Act provides that a dispute can be referred to adjudication "at any time" provided the parties have a contract.

'At any time' can also refer to a dispute after the contract is completed.

What are the requirements within construction contracts?

Section 108 of the Act requires all "construction contracts", as defined by the Act, to include minimum procedural requirements which enable the parties to a contract to give notice at any time of an intention to refer a dispute to an Adjudicator.

The contract must provide a timetable so that the Adjudicator can be appointed, and the dispute referred, within seven days of the notice. The Adjudicator is required to reach a decision within twenty-eight days of the referral, plus any agreed extension, and must act impartially. In reaching a decision an Adjudicator has wide powers to take the initiative to ascertain the facts and law related to the dispute.

The Housing Grants, Construction and Regeneration Act 1996 defines adjudication as:

> *a summary non-judicial dispute resolution procedure that leads to a decision by an independent person that is, unless otherwise agreed, binding upon the parties for the duration of the contract, but which may subsequently be reviewed by means of arbitration, litigation or by agreement.*

In that sense, adjudication does not necessarily achieve final settlement of a dispute because either of the parties has the right to have the same dispute heard afresh in court, or where the contract specifies arbitration.

However, experience since the Housing Grants, Construction and Regeneration Act 1996 came into force shows that the majority of adjudication decisions are accepted by the parties as the final result.

For Option W2, there is no Dispute Reference Table as the Parties have a right to refer a dispute to each other and to the Adjudicator at any time. Any dispute arising under the contract is first referred by a Party issuing a notice to the Senior Representatives and copied to the other Party and the Project Manager stating the nature of the dispute it wishes them to resolve. Each Party then submits their statement of case (no more than ten sides of A4 paper) with supporting evidence.

The Senior Representatives then try to resolve the dispute within three weeks, using any procedure they wish to. At the end of the three weeks, a list is issued showing the issues agreed, and the issues not agreed. The Project Manager and the Contractor put the agreed issues into effect.

The requirements of the Housing Grants, Construction and Regeneration Act 1996 for adjudication to be possible at any time means that the involvement of the Senior Representatives may be bypassed by parties and the dispute referred straight to the adjudicator.

Again, as with Option W1, the Adjudicator is appointed under the NEC4 Dispute Resolution Service Contract. The Adjudicator acts impartially, and if it resigns or is unable to act, the Parties jointly appoint a new Adjudicator. The referring Party obtains a copy and completes the Dispute Resolution Service Contract. If the Parties have not appointed an Adjudicator, either party may ask the nominating body to choose an Adjudicator.

A Party may first give a notice of adjudication to the other party with a brief description of the dispute, details of where and when the dispute has arisen, and the nature of the redress which is sought.

The Adjudicator may be named in the contract in which case the party sends a copy of the notice to the Adjudicator. The Adjudicator must confirm within three days of receipt of the notice that it is able to decide the dispute, or if it is unable to decide the dispute.

Within seven days of the issue of the notice of adjudication, the party:

- refers the dispute to the Adjudicator;
- provides the Adjudicator with the information on which it relies, together with supporting information;
- provides a copy of the information and supporting documents to the other party.

Again, a dispute under a subcontract, which is also a dispute under the ECC, may be referred to the Adjudicator at the same time and the Adjudicator can decide the two disputes together.

The Adjudicator may review and revise any action or inaction of the Project Manager, take the initiative in ascertaining the facts, and the law relating to the dispute, instruct a party to provide further information and instruct a party to take any other action which it considers necessary to reach its decision.

All communications between a Party and the Adjudicator must be communicated to the other party at the same time.

The Adjudicator decides the dispute and notifies the parties and the Project Manager within twenty-eight days of the dispute being referred.

This twenty-eight-day period may be extended by fourteen days with the consent of the referring Party or by any other period by joint agreement between the parties.

Until this decision has been communicated, the parties proceed as if the matter in dispute was not disputed.

The Adjudicator's decision is temporarily binding on the parties unless and until revised by a tribunal and is enforceable as a contractual obligation on the parties.

The Adjudicator's decision is final and binding if neither party has notified the Adjudicator that they are dissatisfied with an Adjudicator's decision within the time stated in the contract and intends to refer the matter to the tribunal.

The Adjudicator may, within five days of giving its decision to the parties, correct any clerical mistake or ambiguity.

Review by the tribunal

A dispute cannot be referred to the tribunal unless it has first been referred to the Adjudicator.

Again, the tribunal may be named by the Client within Contract Data Part 1, normally litigation or arbitration. If arbitration is chosen, the Client must also state in Contract Data Part 1, the procedure, the place where the arbitration is to be held, and the person or organisation who will choose an Arbitrator if the Parties cannot agree a choice, or if the named procedure does not state who selects the Arbitrator.

A party can, following the adjudication, notify the other party within four weeks of the Adjudicator's decision that it is dissatisfied. This is a time-barred right, as the dissatisfied Party cannot refer the dispute to the tribunal unless it is notified within four weeks of the Adjudicator's decision; failure to do so will make the Adjudicator's decision final and binding.

The tribunal settles the dispute and has the power to reconsider any decision of the Adjudicator and review and revise any action or inaction of the Project Manager.

It is important to note that the tribunal is not a direct appeal against the Adjudicator's decision; the parties have the opportunity to present further information or evidence that was not originally presented to the Adjudicator, and also the Adjudicator cannot be called as a witness.

10.13 NEC4 Dispute Resolution Service Contract

The NEC4 Dispute Resolution Service Contract replaced the NEC3 Adjudicator's Contract.

The first edition of the NEC Adjudicator's Contract was published in 1994 and was written for the appointment of an Adjudicator for any contract under the NEC family of standard contracts. The second edition, published in 1998, contained some changes including the need to harmonise with the NEC standard contracts and further editions which had been issued since 1994. The third edition harmonised the contract with the NEC3 family of contracts using either Option W1 or W2.

The Dispute Resolution Service Contract is made up of five parts:

1 General
2 Adjudication
3 Dispute Avoidance Board
4 Payment
5 Termination.

The Dispute Resolution Service Contract can also be used with contracts other than NEC4, though dependent on the contract and the applicable law, some amendment may be necessary.

In the UK, the Housing Grants, Construction and Regeneration Act 1996 has made adjudication mandatory as a means of resolving disputes in construction contracts which fall under the Act. Parties to a contract which does not provide for adjudication as required by the Act have a right to adjudication under the Scheme for Construction Contracts. Schemes which are substantially similar have been published for England and Wales, Scotland and Northern Ireland.

The agreement is between the two disputing parties and the Dispute Resolver, who may also be acting as the Adjudicator.

The contract contains five sections:

1 General
     This covers the obligation upon the Dispute Resolver to act impartially, and to notify the parties as soon as it becomes aware of any matter which could present a conflict of interest.
     There is a definition of expenses, which includes printing costs, postage, travel, accommodation and the cost of any assistance with the adjudication.
     All communications must be in a form which can be read, copied and recorded. This is a requirement of all NEC contracts, but also of the adjudication process itself, with both Parties and the Dispute Resolver having to be copied into any communication between the parties.
2 Adjudication
     This clause only applies if the Dispute Resolver is acting as an Adjudicator.
     The Dispute Resolver cannot decide any dispute that is the same or substantially the same as its predecessor decided. It must make its decision and notify the Parties in accordance with the contract, and in reaching its decisions it can obtain assistance from others, but must advise the Parties before doing so, and also provide the Parties with a copy of anything produced by the assisting party, so that the Parties can be invited to comment.
     The Dispute Resolver's decision is to remain confidential between the parties, and following its decision the Dispute Resolver retains documents provided to him for the period of retention, which is stated in the Contract Data.

An invoice is issued by the Dispute Resolver:

- each time a dispute is referred if an advance payment is stated in the Contract Date;
- after a decision had been notified to the Parties;
- after a termination.

3 Dispute Avoidance Board

This clause only applies if the Dispute Resolver is acting as a Dispute Avoidance Board member.

The Dispute Resolver carries out the duties of a Dispute Avoidance Board member in accordance with the contract between the Parties and collaborates with other members of the Dispute Avoidance Board.

4 Payment

The Dispute Resolver's hourly fee is stated in the Contract Data. There is provision for an advance payment to be made by the referring party to the Dispute Resolver if stated in the Contract Data.

Following its decision, the Dispute Resolver submits an invoice to each Party for their share of the amount due, including expenses. The Parties pay the amount due within three weeks of receiving the Dispute Resolver's invoice.

Interest is paid at the rate stated in the Contract Data on any late payment. The Parties are jointly and severally liable to pay the amount due to the Dispute Resolver; therefore, if one party fails to pay, the other must pay including any interest on the overdue amount, and recover the amount from the other Party.

5 Termination

The Parties may, by agreement, terminate the Dispute Resolver's agreement.

Also, the Dispute Resolver may terminate the agreement if there is a conflict of interest, it is unable to decide a dispute, an advance payment has not been made, or it has not been paid an amount due within five weeks of the date by which the payment should have been made. The Dispute Resolver's appointment terminates on the date stated in the Contract Data.

10.14 The tribunal

Whilst Contract Data Part 2 requires the Client to enter the form of the tribunal, it does not say what choices are available. Other contracts refer disputes to expert determination or Dispute Review Boards, but under the ECC the Client normally selects arbitration or litigation. Later, within the Contract Data the Client is required to state the arbitration procedure, the place where the arbitration is to be held, and who will choose an Arbitrator.

There are a number of differences between the two but essentially arbitration is hearing the dispute in private and litigation is hearing the dispute in a public court.

It must be stressed that the role and authority of the tribunal is not to review and appeal against the previous decision of an Adjudicator, but to resolve the dispute independent of the adjudication; therefore, it is quite in order for parties to introduce new evidence or materials to the tribunal.

10.15 Arbitration

Any dispute between two or more parties can be resolved through the courts, a process known as litigation.

However, litigation has many disadvantages, not least the cost of a full court hearing, but in some cases a lengthy waiting time before the matter actually reaches the courts, so arbitration has become increasingly used as a simpler, more convenient method of dispute resolution.

Arbitration is the hearing of a dispute by a third party, who is often not a lawyer, though that term has a very wide meaning, but is an expert in the field in which the dispute is based, and can give a decision based on opinion, which is then legally binding on both parties.

Arbitration is not a new concept, in fact it has been in existence for almost as long as the law itself, the first official recognition in the UK being the Arbitration Act 1697, which largely governed disputes about the sale of livestock in cattle markets. With the continuing growth of freight transport by ship, and the advent of railways and other forms of transport, a large number of arbitration cases ensued, and as a result Parliament passed the Arbitration Act 1889.

Various amendments have been introduced over the years which followed with the current legislation being the Arbitration Act 1950, as amended by the Arbitration Acts of 1975 and 1979. The Arbitration Act 1996 is the current legislation related to arbitration in England and Wales.

One does not issue writs in arbitration as one would with litigation; *both* parties agree to enter into arbitration as a pre-agreed provision within a contract and thus be bound by the decision of the Arbitrator.

Arbitrators are appointed by one of THREE methods:

(i) As a result of a pre-agreed decision between the parties, normally though a provision within a contract to go to arbitration in the event of a dispute.
     In this case the Arbitrator itself may be named or subject to nomination by a professional body such as the Chartered Institute of Arbitrators.
(ii) By Statute, legislation in many countries includes the provision for disputes to be settled by arbitration.
(iii) By Order of a Court.

The Arbitration Act is fairly brief and sets out procedures in the absence of any agreement to the contrary between the parties.

As arbitration is a more flexible arrangement than litigation, whatever procedure the Arbitrator and both parties agree is sufficient in an arbitration case.

Whilst most forms of contract provide for resolution of disputes by arbitration, it must be stressed that it should only be seen as a last resort when all attempts at negotiation and other resolution methods have failed.

It may be surprising to hear that arbitration is common as a settlement procedure for disputes under trade agreements, maritime and insurance, consumer matters such as package holidays, construction industry disputes and property valuations.

The contracting party that initiated that the dispute be referred to arbitration is always referred to as the 'Claimant', whilst the other party is referred to as the 'Respondent'. Should there be a Joint Agreement to go to arbitration, the Arbitrator will decide who is the Claimant, and who is the Respondent.

It is important that these titles are clarified, as the Arbitration Rules repeatedly refer to them.

There are several advantages in using arbitration rather than litigation:

(i)   Privacy

This is a distinct advantage, as in litigation, private and often personal matters are debated in full view of the public in open court, and often published in the press and technical magazines, whereas arbitration is carried out in private sessions, usually at the Arbitrator's office or at a pre-arranged venue.

(ii) Convenience

Arbitration is carried out at a time and place to suit the parties, for example evenings and weekends, not when the court is available for session.

(iii) Speed

The dispute is settled efficiently without the normal delays and procedural matters involved in often fairly lengthy and formal litigation.

(iv) Simplicity

There are fewer technical procedures than in the courts, and the parties set their own procedures within a set of ground rules.

(v)   Expertness

The Arbitrator will be an expert on the matter in dispute, for example a structural engineer where the dispute is about structural design issues, whereas judges and other legal professionals are experts only on principles of law. The dispute will often not hinge on a legal problem, for example in matters of design, who better than a designer to resolve it?

(vi) Thoroughness

Because the Arbitrator is a skilled technical person, the dispute can often be dealt with more thoroughly and matters more rigorously aired than in a confined and formal court atmosphere.

(vii) Expense

It can be readily seen that arbitration is generally a less expensive option than litigation, though it must not be seen as a cheap way to resolve a dispute. In some complex cases arbitration can be even more expensive than litigation.

It is worth mentioning at this point that, in many countries, legal aid, which can be available for litigation, is not available for arbitration.

The main disadvantages of using arbitration rather than litigation are:

(i)   Lack of legal knowledge

A dispute may hinge on a difficult point of law, which the Arbitrator has insufficient legal knowledge to resolve, as it is an expert in the field relating to the dispute, not in the law itself. It is, however, entitled to seek legal advice on a point of law, but not on the dispute itself, and also the disputing parties can refer a question of law to the High Court.

N.B. The arbitration award itself cannot be referred to the Courts unless both parties consent, or the Courts are referred to on a point of law, the resolution of which could substantially overturn an Arbitrator's decision.

Further, Arbitrators cannot be sued for negligence as they do not hold themselves up as experts in the sense of being advisory bodies and therefore liable for giving wrongful advice.

(ii)   Precedents

Arbitration is not subject to rulings in previous cases as with the courts, each case being judged on its own merits on the day, and also to some extent subject to the personality of the Arbitrator, therefore there is no guide as to how successful an action could be.

It is prudent to always seek professional advice before resorting to arbitration, as whilst it is simpler than Court action the Arbitrator's decision is final and binding, therefore a party must be confident that they have a strong case.

(iii)   Decision making

It has been said that the Arbitrator, being a technical person and possibly having been involved in a similar dispute themselves, tends to be fairer than judges in litigation; they also tend to sympathise with both parties and can split the difference between the parties when making their awards as they have possibly been in a similar position themselves, but judges are inclined to take colder decisions and award wholly in favour of one of the parties. It has to be said that this is not a commonly held view!

Reference to arbitration

Although arbitration is a different process to adjudication, the same initial principles apply before arbitration can commence, in that there must be a dispute, there must be an agreement to arbitrate, which will normally be provided for within the contract, and thirdly a party must submit a notice to initiate the process.

Proceedings begin by either party sending a written notice to the other stating that they wish to refer the dispute to arbitration; the form and process may be dictated by the arbitration procedure stated by the Client within Contract Data Part 1.

The parties may agree on a person to act as Arbitrator, or a professional body will select and appoint the Arbitrator. In some cases, where the parties cannot agree a choice and there is no provision for a third party to do it, a court may select an Arbitrator. There is normally only one Arbitrator, but in some cases three, consisting of two Arbitrators and a third who acts as an umpire. Any correspondence flowing between the two parties as from the date of appointment must then also be copied to the Arbitrator(s).

The next step is that the Arbitrator writes to the parties accepting the appointment, then a Preliminary Meeting is held between the parties at a time and venue to be directed by the Arbitrator to decide the form of the hearing, which may be on a documents-only basis or an oral hearing. Documents-only tends to be cleaner and quicker as there is less opportunity for the parties to debate the issue.

The role of the Arbitrator

The role of the Arbitrator is only to examine the evidence before him enabling him to resolve the dispute, and nothing more; it cannot look into other matters not in dispute. It can conduct the arbitration as it wishes, though it must gear the procedures in a way that is not detrimental to either party.

It must also have a regard for the rules of evidence.

For example:

1 Hearsay evidence is not permissible.
2 It is not permitted to use evidence from someone not qualified to give such evidence.
3 The Arbitrator is not allowed to take into account any evidence it has discovered itself, which was not provided by either party.
4 It is not allowed to use any privileged evidence, for example that which is marked "Without Prejudice".
5 In addition, the rules of discovery apply where either party is entitled to have access to the file or documents of the other provided that it relates to matters at issue.

Inspection by the Arbitrator

The Arbitrator is entitled to inspect any property, work or material at any premises, and may invite the Claimant and/or the Respondent to accompany him purely for the purpose of identifying the work or materials.

No-one else is allowed to accompany the Arbitrator unless it specifically invites him.

The award

The award is binding upon both parties and must be served in the appropriate manner with full headings, summary of issues in dispute, outline of events leading to the dispute, its decision based on the hearing and finally the monetary award.

There is very limited scope for appeal against an Arbitrator's award. Usually, appeals can only be based on a claim that the Arbitrator was wrong on a point of law, or there was serious and proven irregularity.

Arbitration awards can also be enforced in court if necessary.

Arbitrator's fees and expenses

Both parties will normally incur costs, though it is the Arbitrator's decision as to how the costs are apportioned. Both parties are initially liable to the Arbitrator for payment of its fees and expenses, but the Arbitrator will direct the parties as to the proportion they are liable for once the decision is reached.

Should the parties agree on a settlement of the dispute at any time during the arbitration, they will jointly be liable for the Arbitrator's costs.

10.16 Litigation

It is not proposed to go into any detail about the litigation process from initiation to judgement by a court, first as the contract does not give any details of the process, but also the proceedings themselves will depend on the law of the contract, which in turn is dependent on the location of the project.

However, the process will normally involve the Claimant's representative, usually a Solicitor issuing a claim form or writ to the appropriate court; papers are then served on the Defendant. The Defendant then replies to the service by either admitting or denying the claim against him; if he denies it, he replies by sending his defence which may also be accompanied by any counter-claim. The case is then allocated to a court. There then follows a period of accumulation, collation and presentation of evidence, and hearings, again the process, the submission and the form of the hearing being dependent upon the law of the contract, until the court passes judgement. Cases in most UK courts will normally see barristers acting as advocates on behalf of the parties alongside the solicitors who have advised the parties.

10.17 Option W3

Published construction contracts in the past have usually included some form of dispute resolution provisions including legal proceedings, arbitration, dispute adjudication boards, adjudication (by contract and/or by statute), expert determination and mediation. In the absence of such provisions, or a mutual agreement between the parties to adopt their own methods, the default would be for the parties to refer to legal proceedings.

In recent years, contracting parties, and in turn contract drafters, have also looked towards dispute avoidance to try to prevent issues escalating into dispute in the first place. This can be in the form of more balanced contract conditions, partnering agreements, target contracts, but also by including specific dispute avoidance provisions.

Option W3 within the NEC3 ECC is such a provision, and is a new addition to the NEC4 contracts, which is only included in some of the contracts, for example the ECC. Note that W3 is an Option for avoiding a dispute, not resolving it.

Option W3 is used when a Dispute Avoidance Board is the method of dispute resolution and the United Kingdom Housing Grants Constriction and Regeneration Act 1996 does not apply.

The Dispute Avoidance Board consists of one or three members (odd number) as identified within the Contract Data Part 1. If the Contract Data states three members, the Parties jointly choose the third member.

The Dispute Avoidance Board is appointed using the NEC4 Dispute Resolution Service Contract (see above).

If a member of the Dispute Avoidance Board is not named within the Contract Data or is unable to act, then the Parties jointly choose a replacement. If the Parties do not do so, the Dispute Avoidance Board may do so themselves within seven days of the request to do so.

The Contract Data states the frequency of visits to Site by the Dispute Avoidance Board from the starting date to the defects date, but may make further visits if requested by the Parties.

The members of the Dispute Avoidance Board, their employees or agents, are not liable for any action or failure to take action to resolve a potential dispute, unless the failure was in bad faith.

A potential dispute under the contract may be referred to the Dispute Avoidance Board, two to four weeks after notification to the other Party and to the Project Manager.

The Parties are required to make all relevant information available to the Dispute Avoidance Board, who visits the Site (where applicable) and helps the Parties to settle the potential dispute without formal referral as a dispute.

Review by the tribunal

A dispute cannot be referred to the tribunal unless it has first been referred to the Dispute Avoidance Board as a potential dispute.

A party can, following the Dispute Avoidance Board's recommendation, notify the other Party that it is dissatisfied. This is a time-barred right, as the dissatisfied Party cannot refer the dispute to the tribunal unless it is notified within four weeks of the Dispute Avoidance Board's recommendation.

The tribunal settles the dispute and has the power to reconsider any recommendation of the Dispute Avoidance Board and review and revise any action or inaction of the Project Manager.

It is important to note that the tribunal is not a direct appeal against the Dispute Avoidance Board's recommendation; the parties have the opportunity to present further information or evidence that was not originally presented to the Dispute Avoidance Board, and also the Dispute Avoidance Board cannot be called as a witness.

11 Tenders

11.1 Deciding the procurement strategy

Procurement strategy considers the optimum way of achieving the client's objectives, taking into account its current and future needs and the inherent risks in achieving those objectives.

A well-considered and appropriate procurement strategy together with a clear, robust and unambiguous tender document is the key to avoiding or minimising contractual problems at a later date.

Deciding the procurement strategy is the foundation stone to a successful outcome to a contract, but this process is very often overlooked as clients and their advisers, in their urgency to get projects out to tender and commenced on site, tend to favour the procurement strategy most familiar to them, rather than to spend time properly considering the project in terms of the client's resources and its objectives in terms of time, cost and quality, with the additional factor of risk having to also be considered and evaluated.

(i) Time

Most clients want their projects to be completed as quickly as possible. This is particularly relevant in the retail sector where the client needs to open a shop as early as possible in order to sell goods. Some projects also have an absolute deadline, e.g. international sporting events.

If the client sees time as being the most important factor, it may have to pay more money, and quality may lower in order to achieve that. The procurement method will have to consider timing through programme requirements, but also through provisions such as delay damages.

(ii) Cost

All clients will have some form of budget, so this will always be a major consideration. The budget may be fixed, representing all that the client has available through its own funds or borrowings in order to fund the project, or may have some degree of flexibility. The form of procurement will need to address the budget and any inherent flexibility.

(iii) Quality

Finally, quality will always be a major factor, though if the above two are more important, some clients will lower the expected standards. It is important that the design lowers the standards, not that a "cheap" contractor is employed who will unilaterally lower the standards itself!

It is rarely achievable for all three to be fully satisfied; it is usually a balance.

Once these have been established and analysed, if using the ECC, the appropriate decisions can be made on Main and Secondary Options, whether the contractor is to design all, or portions of the work, but also the allocation of responsibilities and the timing for preparing the relevant documents.

Clients should always carefully select a procurement strategy most appropriate to their needs, the project and the various risks involved. The procurement strategy should consider value for money in a holistic way, considering life cycle costing as well as the capital cost of building the project, though it is not intended to consider life cycle costs in any detail within this book.

The procurement strategy then leads to a contract strategy which should be capable of achieving the aims of the procurement strategy.

The initial briefing is the fundamental first step in establishing what is required, when it is required, and what the client is proposing, or can afford to pay.

Through a process of interviewing key members of the client's team and reviewing its business plans, specific objectives, options and constraints, a procurement strategy is developed which forms the basis for future assessments and recommendations.

Whilst it may sound obvious, it is vital that the client understands what it wants and can convey it clearly. It may not know what it wants, but also it can often be swayed by consultants or other third parties.

Whilst many traditional contracts are based on bills of quantities with the client designing the project, and detailed bills of quantities being sent to each tendering contractor for pricing, in the past 20 years there has been a movement away from the use of traditional bills of quantities and towards contractor-designed projects with payment mechanisms such as milestone payments and activity schedules, with payment based on progress achieved, rather than quantity of work done.

This contractor-designed growth has given rise to the increasing use of various design and build contracts, performance specifications and design, build, finance and operate projects.

There are a number of reasons for allocating some or most of the design to the contractor (see also Chapter 2):

- the design and construction periods can overlap, leading to faster delivery of the project;
- the contractor can utilise its experience and preferred methods of construction to build rationally designed projects which minimise costs and programme durations;
- the temporary works/permanent works interface and influence of design is rationalised as both remain with the contractor and therefore the permanent works should be more "buildable";

- the traditional design/construction interface and the risks associated with it are transferred to the contractor;
- the management of the design risk by the contractor can result in greater certainty of the time, cost and performance and project objectives being met.

There has also been an increasing use in the past 15 years of target contracts which have been encouraged by the increasing use of partnering and collaborative working arrangements, and the associated more equitable sharing of risk.

Through its Main and Secondary Options, the ECC caters for all the procurement methods commonly used in construction contracts throughout the world.

Factors to take into account in deciding which procurement option to use, and in turn which Main and Secondary Options to use with the ECC, include the following:

- How important to the client is certainty of price, particularly if changes are expected? Lump sum options will give greater certainty of price.
- What resources and expertise does the client have?
- Which party is to be responsible for design and/or which party has the necessary design expertise? Design can be assigned to the contractor under any of the main options.
- How important to the client is an early start and/or early completion? Early commencement and early completion may lead the client toward a cost-reimbursable solution as lump sum and remeasurement procurement options require a considerable time to prepare the tender documentation for pricing.
- How clearly defined is the scope of the works, and what form does it take, for example is it in the form of detailed drawings and specification, or performance-based criteria? Again, the contracts can be used with detailed drawings or performance-based criteria, where the contractor carries out the design.
- What is the likelihood of change to those defined requirements? Many clients prefer a remeasurement option where there is likelihood of many changes.
- What is the client's attitude toward risk? Who is best placed to manage the inherent risks within the project? Lump sum contracts assign greatest risk to the contractor, cost-reimbursable contracts assign greater risk to the client. It is a poorly acknowledged fact that the client always pays for risk, regardless of whether it is carried within the contract by the client or the contractor. If it is the contractor's risk it is deemed to have priced and programmed for it; if it is a client's risk it will pay for it if it occurs.

11.2 Partnering arrangements

The past 20 to 25 years, particularly since the publication of the Latham and Egan reports, have seen the growth of partnering and framework agreements.

The US Construction Industry Institute defines partnering as:

A long-term commitment between two or more organisations for the purpose of achieving specific business objectives by maximising the effectiveness of each participant's resources . . . the relationship is based upon trust, dedication to common goals and an understanding of each other's individual expectations and values.

Partnering is a medium- to long-term relationship between contracting parties, whereby the contractor and in turn consultants and various other parties are not required to tender for each project, but are awarded the work, normally by negotiation.

Advantages of partnering agreements are:

- no rebidding for future projects;
- developed relationships based on trust;
- contractors are appointed earlier and can contribute to the design and procurement process;
- greater cost certainty;
- continuous improvement by transferring learning from one project to another;
- improved working relationships;
- continuous workflow;
- speed of procurement.

The construction industry is known to be a high-risk business, and many projects can suffer unexpected cost and time overruns, frequently resulting in disputes between the parties.

The risks within a project are initially owned by the client, who may choose to adopt a "risk transfer" approach where the risks are assigned through the contract to the contractor who has the opportunity to price and programme for them, or a "risk embrace" approach where the client retains the risks. In reality, most contracts are a combination of the two. The traditional approach to risk management is that of risk transfer, which is fine if the scope of work is clear and well defined, however, in recent years clients have become more aware that they can achieve their objectives better by adopting a more "old fashioned" risk embrace culture.

11.3 Provisional & Prime Cost (PC) Sums

Provisional & PC Sums are used in contracts where there are elements of work which are not designed or cannot be sufficiently defined at the time of tender and therefore a sum of money is included by the client within the bill of quantities or other pricing document to cover the item. When the item is defined or able to be properly defined the contractor is given the information which allows him to price it, the Provisional Sum or PC Sum is omitted, and the price included in its stead.

The problem with Provisional Sums and PC Sums is that they reduce the competition amongst tenderers as they are not priced at tender stage, and also with Provisional Sums if they are "defined" the contractor is deemed to have allowed time in its programme for them; if they are "undefined", it has not.

The NEC contracts do not provide for Provisional Sums or PC Sums, the principle being that:

> The Client decides what it wants at tender stage so it can be designed and accurately described and the tendering contractors can properly price and programme for them,

or

> When the Client decides what it wants, the Project Manager can give an instruction to the Contractor which changes the Scope; it is a compensation event under Clause 60.1(1) and the time and/or cost effect can be priced at the time. In that case, the cost of the work is held in the scheme rather than in the contract.

11.4 The pre-tender planning stage

The contractor selection process normally follows the following steps:

- Compiling the shortlist of tenderers
- Criteria for contractor selection
- Information for contract tender enquiries
- Issue tender documents
- Receipt of tenders
- Comparison of tenders
- Contract award

Each of the above stages will be dealt with in the form of a checklist.

11.5 Preparing tender documents

The tender documents for an ECC contract will normally consist of:

(i)   Invitation to Tender (including instructions to tenderers on time and place of tender submission)
(ii) Form of Tender
(iii) The Pricing Document, i.e. Bills of Quantities for an Option B or D contract, possibly a pro forma activity schedule for an Option A or C contract
(iv) Scope
(v)   Site Information

(vi)   Contract Data Part 1 (completed by the client)
(vii) Contract Data Part 2 (blank pro forma to be completed by tendering contractors)

In addition there may be other information, dependent on the applicable legislation, for example in the UK, where the Construction (Design & Management) Regulations apply, Pre Construction Information would be included within the tender documents.

Invitation to tender

As soon as it has been decided to select a contractor by competitive tender, a shortlist of those considered suitable to be invited to tender should be compiled.

It is often the case that clients and their representatives already hold a list of contractors of established skill, integrity, responsibility and proven competence. This should be considered as a matter of policy within the client's organisation.

Clients should review their lists periodically so as to exclude firms whose performance has been unsatisfactory and to allow the introduction of new firms.

The cost of preparing tenders is a significant element of the overheads both of the main contractor and in turn its subcontractors. Tender lists therefore should be kept as short as practicable. Enquiries should be kept to between three and six, depending upon the type of work and size of project.

The use of a register of contractors at the selection stage is recommended. This may lead to establishing contractor record cards relating to performance criteria on previous contracts.

It is important that such records are regularly updated and consideration given to adding new contractors to the register. It may be advisable at this stage to consider using a computer database for compiling the register. This may prove advantageous to a medium- to large-sized organisation in order to speedily access contractor information.

It is important that close relationships are encouraged between the client and the contractor, and in turn its subcontractors for the survival of both parties in the long term.

Information for tender enquiries

It is critical that clients properly identify what information should be contained in the tender enquiry to contractors.

Checklist for invitation to tender

- Details of Main Contract works
- Job title and location of site
- Name of Client
- Names of Consultants

- Details of form of contract
- General description of the works
- Bills of quantities (if applicable)
- Copies of relevant drawings
- Specification
- Details of where further documents may be inspected
- Time period for completion of work (if known)
- Names of adjudicator (in case of dispute)
- Contractor's responsibility for site arrangements and facilities:
  - Watching and lighting
  - Storage facilities
  - Unloading, hoisting and getting in materials
  - Scaffolding
  - Water and temporary electrical supplies
  - Safety, health and welfare provisions
  - Licences and permits
  - Any additional facilities

Checklist for contractor selection

- Previous experience with the contractor.
- The contractor's ability to manage its resources and liaise with the client's staff. Good relationships between parties are an essential requirement to developing a team approach to a successful project.
- Financial standing of the contractor.
- The contractor's expertise which it can bring to the project.
- The contractor's reputation and its standing with the client.
- The current commitment of the contractor's organisation.
- Their current workload with other clients should be determined and serious consideration given to their ability to cope with the increased work. A large number of contractors just cannot say "no" when it comes to taking on more work.
- The acceptability of the contractor to the client. On many contracts the contractor is required to name its subcontractors at the tender stage.
- The competitiveness of the contractor's price. The price must be right otherwise the contractor will never win any work. Price discounts which may be applicable and the contractor's response to negotiation may be an important factor.
- The contractual risk which the contractor is expected to take.
- The ability of the contract organisation to meet "quality assurance criteria" as laid down by the contract or as specified by the client.
- References available from the contractor. These include trade and bank references.
- The willingness of the contractor to allow previous contract work to be inspected.

- It is important that good relationships are established between the client and the contractor as early in the planning process as possible. This is especially important where the main contractor intends to sublet all the work on a particular project.

N.B. There is no "pro forma" version of "Instructions to Tenderers" within the NEC contracts.

Any invitation to tender will give a brief description of the work, who it is for, and how tenders may be submitted.

Other inclusions will cover such matters as:

- Time, date and place for the delivery of tenders.
- What documents must be included with the tender, including a programme.
- The policy regarding on alternative and/or non-compliant bids.
- Arrangements for visiting the site including contact details.
- Rules on non-compliant bids.
- Anti-collusion certificate.

Form of Tender

There is no "pro forma" Form of Tender within the NEC contracts. This is the tenderer's written offer to execute the work in accordance with the contract documents.

The Form of Tender is normally in the form of a letter with blank spaces for tenderers to insert their name and other particulars, total tender price, and other particulars of their offer. It is essential to have a standard Form of Tender so that all tenderers utilise the same form, to make comparison of tenders easier.

The pricing document

When using the ECC, the pricing document will depend on the Main Option chosen.

With Options B or D, the client is required to send the tenderers a Bill of Quantities. This comprises a list of items giving the quantities and a description of the work included in the contract and, in conjunction with the other contract documents, forms the basis upon which tenders are obtained and on which the contractor is paid. The Bill of Quantities is compiled in accordance with a Standard Method of Measurement and must not conflict with the Scope.

Scope

Scope is defined by Clause 11.2(16) as

"information which

- specifies and describes the works or
- states any constraints on how the contractor Provides the Works

and is either

- in the documents which the Contract Data states it is in or
- in an instruction given in accordance with this contract."

The Client provides the Scope and refers to it in Contract Data Part 1; if the Contractor provides the Scope, e.g. for its design, this is included in Contract Data Part 2.

The main documents within the Scope are normally the drawings and specifications, but will also include:

Description of the works

- a statement describing the scope of the works;
- schematic layouts, plan, elevation and section drawings, detailed working and/or production drawings (if relevant and available), etc.
- a statement of any constraints on how the contractor Provides the Works, e.g. restrictions on access, sequencing or phasing of works, security issues.

Plant and Materials

- materials and workmanship specifications;
- requirements for delivery and storage;
- future provision of spares, maintenance requirements, etc.

Health and safety

- Specific health and safety requirements for the site which the contractor must comply with, particularly if the site is within existing premises, including house safety rules, evacuation procedures, etc.
- Any pre-construction information and health and safety plans for the project.

Financial records

- Details of the accounts and records to be kept by the contractor.

Contractor's design

- The default position is that all design will be carried out by the client. If any or all of the design is to be carried out by the contractor, it should be included within the Scope together with any performance requirements to be met within the contractor's design. In addition any warranty requirements and also any future novation requirements should be included within the Scope;
- Any design acceptance procedures should be included, including timescales for submission and acceptance of design.

Completion

- Completion is defined under Clause 11.2(2), but the Scope should also define any specific requirements which are required in order for completion to have taken place for example, and requirement for:
  - o "As built" drawings
  - o Maintenance manuals
  - o Training documentation
  - o Test certificates
  - o Statutory requirements and/or certification.

Services

- Details of other contractors and Others who will be occupying the Working Areas during the contract period and any sharing requirements.

Subcontracting

- Lists of acceptable subcontractors for specific tasks;
- Statement of any work which should not be subcontracted;
- Statement of any work which is required to be subcontracted.

Programme

- Any information which the contractor is required to include in the programme in addition to the information shown in Clause 31.2. Also, if the client requires the contractor to produce a certain type of programme or to submit it using a certain brand of software, then this should be clearly detailed within the Scope.

Tests

- Description of tests to be carried out by the contractor, the supervisor and others including those which must be done before Completion;
- Specification of materials, facilities and samples to be provided by the contractor and the client for tests;
- Specification of Plant and Materials which are to be inspected or tested before delivery to the Working Areas;
- Definition of tests of plant and materials outside the Working Areas which have to be passed before marking by the supervisor.

Title

- Statement of any materials arising from excavation or demolition to which the client will have title;
- Requirements for plant and material to be marked.

Others

- There are also certain specific requirements for statements to be made in the Scope from certain main and secondary options in the conditions of contract.

The Scope must be carefully drafted in order to define clearly what is expected of the contractor in the performance of the contract and therefore included in the quoted tender amount and programme. If the contract does not cover all aspects of the work, either specifically or by implication, that aspect may be deemed to be excluded from the contract.

A comprehensive, all embracing description should therefore be considered for the scope of work clause, which should be supplemented by specific detailed requirements. If reliance is placed solely on a very detailed scope description an item may be missed from this detailed description and be the subject of later contention.

Where items of equipment are to be fabricated or manufactured off site by others, it is advisable that the contract sets out the corresponding obligations and liabilities of the respective parties, particularly if these are to form an integral or key part of the completed works.

The Scope describes clear boundaries for the work to be undertaken by the contractor. It may also outline the client's objectives and explain why the work is being undertaken and how it is intended to be used. It says what is to be done (and maybe what is not included) in general terms, but not how to do it or the standards to be achieved. It explains the limits, where the work is to interface with other existing or proposed facilities. It may draw attention to any work or materials to be provided by the client or others. It should also emphasise any unusual features of the work or contract, which tenderers might otherwise overlook.

This is the document that a tenderer can look to, to gain a broad understanding of the scale and complexity of the job and be able to judge its capacity to undertake it. It is written specifically for each contract. In some respects, it is analogous to a shopping list. It should be comprehensive, but it should be made clear that it is not intended to include all the detail, which is contained in the drawings, specifications and schedules.

The Scope will also include drawings, which again should provide clear details of what the contractor has to do. Clearly, tenderers must be given sufficient information to enable them to understand what is required and thus submit considered and accurate tenders.

It will also include the specification which is a written technical description of the standards and various criteria required for the work, and should complement the drawings. The Specification describes the character and quality of materials and workmanship, for work to be executed.

Again, it may lay down the order in which various portions of the work are to be executed. As far as possible, it should describe the outcomes required, rather than how to achieve them. It is customary to divide the work into discrete sections or trades (e.g. drainage, concrete, pavements, fencing, etc), with clauses written to cover the materials to be used, the packaging, handling and storage of materials (only if necessary), the method of work to be used (only if necessary), installation criteria, the standards or tests to be satisfied, any specific requirements for completion, etc.

The specification is an integral part of the design. This is often overlooked, with the result that inappropriate or outmoded specifications are selected, or replaced by a few brief notes on the Drawings. The designer should spend an appropriate amount of time specifying the quality of the work, as it is not possible to price, build, test or measure the work correctly unless this is done.

The Scope describes what the contractor has to do in terms of scope and standards, and in some cases must not do or include in order to provide the works. It may also include the order or sequence in which the works are to be carried out. It also details where the work is to interface with other existing or proposed contractors or facilities. It will also include any work or facilities or materials to be provided by the client or others. It may also include any unusual features of the work, which tenderers might otherwise overlook, for example any planning constraints, etc.

This is the document from which a tenderer can gain a broad understanding of the scale and complexity of the job and its capacity to undertake it.

It is written specifically for each contract.

The Scope should be:

- Clear – unambiguous
- Concise – not over wordy
- Complete – have nothing missing

In respect of the contractor's design, a reason for not accepting the design is that it does not comply with either the Scope or the applicable law. Again, if the Scope was lacking and the contractor provided a design that complied with it, and the applicable law, could the contractor escape liability for the defective design? Secondary Option X15 states that "the contractor is not liable for Defects in the works due to its design so far as it proves that it used reasonable skill and care to ensure that its design complied with the Scope". Again, this places a heavy burden on the Scope.

Site Information

Site Information is defined in Clause 11.2(18) as "information which describes the Site and its surroundings" and is identified in Contract Data Part 1.

Site Information may include:

(i)   Ground investigations, borehole and trial pit records and test results. If the client has obtained such information, it should not be withheld from the tenderers, though the tenderers should be aware that the Site Information alone cannot be relied upon in terms of a possible compensation event (see Clause 60.2).
(ii)   Information about existing buildings, structures and plant on or adjacent to the Site.
(iii)   Details of any previously demolished structures and the likelihood of any residual surface and subsurface materials.
(iv)   Reports obtained by the client concerning the physical conditions within the Site or its surroundings. This may include mapping, hydrographical and hydrological information.
(v)   All available information on the topography of the site should be made accessible to tenderers, preferably by being shown on the Drawings.
(vi)   Environmental issues, for example nesting birds and protected species.
(vii)   References to publicly available information.
(viii)   Information from utilities companies and historic records regarding plant, pipes, cables and other services below the surface of the site.

People are often confused as to what is included within the Scope, and what is included within the Site Information.

As a general statement, if the contractor needs to just be aware of something related to the Site, then it should be included within the Site Information.

However, if the contractor has to do some work in connection with what is on Site as part of its obligation to Provide the Works, and therefore it should have the opportunity to price and/or programme it within its tender, then it should be included within the Scope.

This does not mean that the contractor does not price and/or programme anything within the Site Information.

As an example, if there is Japanese Knotweed (Fallopia Japonica) on Site and the contractor needs to be aware of its existence, then it should be included within the Site Information. The contractor will then take measures as required within its temporary works to keep away from it, and if necessary fence off the area. If the contractor needs to remove the Japanese Knotweed, then it should be included within the Scope.

It is vital that care is taken to get the Site Information correct. Tenderers must be given sufficient information to enable them to understand what is required and thus to submit considered and well-priced tenders.

Under Clause 60.3, if there is an ambiguity or inconsistency within the Site Information, the Contractor is assumed to have taken into account the conditions most favourable to doing the work.

Whilst many interpret that as the Contractor allowing the cheapest way of doing the work, it may be the easiest or quickest way.

In the event of the Contractor notifying a compensation event under Clause 60.1(12) for unforeseen physical conditions, it is assumed to have taken into account the Site Information, publicly available information referred to in the Site Information, information obtained from a visual inspection of the site, and other information which an experienced contractor could reasonably be expected to have or to obtain. So the Contractor cannot rely solely on the Site Information in terms of how it prices and programmes the works.

The Contract Data

The Contract Data provides the information required by the conditions of contract specific to a particular contract. Other contracts call this the "Appendices" or "Contract Particulars".

Part 1

Part 1 consists of data provided by the Client, the sections of the Contract Data aligning with the sections of the core clauses.

The Contract Data requires the Client, or the party representing the Client, to identify:

SECTION 1 – GENERAL

- The selected Main and Secondary Options
- A description or title for the works
- The names and contact details of the Client, the Project Manager, the Supervisor and the Adjudicator.

  Whilst the Project Manager and the Supervisor must always be named individuals, many clients choose to insert company names for these parties within the Contract Data and to separately identify the named individuals. It is also not uncommon for clients to name directors or partners of their respective companies, then the authority of the Project Manager and Supervisor is delegated to the individuals under Clause 14.2.
- The Scope and Site Information are also identified within this section, though these are normally incorporated by reference to separate documents rather than listing drawing numbers and specification references within the Contract Data. If this is the case, the separate documents must be clearly defined.
- The boundaries of the site are defined, normally by reference to a specific drawing or map.
- As the NEC4 contracts are intended for use worldwide, the language and the law of the contract are also entered.
- The period for reply is the period that the parties have to reply to submissions, proposals, notifications, etc. where there is no period specifically stated

elsewhere within the contract, for example the period within which a party should reply to an early warning notice, or the Project Manager should reply to the contractor submitting the particulars of its design or the name of a proposed Subcontractor.

Different periods for reply can be inserted into Contract Data Part 1 for different communications, also they can be mutually extended by agreement between the communicating parties.

The period of reply would clearly not apply to the Contractor's submission of a programme for acceptance or the submission of a quotation for a compensation event.

- There is reference to matters to be included within the Early Warning Register and the frequency of early warning meetings.

SECTION 2 – CONTRACTOR'S MAIN RESPONSIBILITIES

- Key Dates and Conditions are inserted (if applicable).
- The frequency of the requirement for the Contractor to prepare forecasts of the total Defined Cost for the whole of the works (if Option C, D, E or F is used).

SECTION 3 – TIME

- The starting date.
- Access date(s).
- The frequency of requirement for the Contractor to submit revised programmes is inserted within this section. Many clients change the reference to "one calendar month" rather than "weeks", to align with monthly progress meetings or reporting requirements.
- The completion date for the whole of the works. If the Client has decided the completion date for the whole of the works, the date is inserted here; alternatively the tendering contractors may be required to insert a date in Contract Data Part 2.
- Whether the Client is prepared to take over the works before the Completion Date.

When the Contractor completes the works, the Project Manager certifies Completion and the Client takes over the works not later than two weeks after Completion, even though takeover may be before the Completion Date. If the Client is not willing to take over the works before the Completion Date, the statement to that effect should remain in the Contract Data; if not it should be deleted. This statement is referred to in Clause 35.1.

- The period after the Contract Date when the Contractor is to submit a first programme for acceptance.

SECTION 4 – TESTING AND DEFECTS

- The period after the Contract Date when the Contractor is to submit a quality policy statement and a quality plan.
- The defects date is identified as a number of weeks after Completion of the whole of the works. This identifies the period the contractor is initially liable for correcting defects.
- The defect correction period is also stated, this being the period within which the contractor must correct each notified defect, failing which the Project Manager assesses the cost to the client of having the Defect corrected by Others.

The Contract Data provides for three entries to be inserted here if required. A defect period correction period for the whole of the works, with two further entries for alternative periods for specific parts of the works, dependent on the urgency to have them corrected, for example:

All works   7 days
except for . . .
Securi ty installations   24 hours
Other Mechanical and electrical installations   48 hours

Clearly any defect that has a potential effect on health and safety should be corrected as soon as possible.

SECTION 5 – PAYMENT

- Again, as the NEC3 contracts are intended for use worldwide, the currency of the contract is entered.
- The assessment interval is also entered, which again is often expressed as "one calendar month" rather than in "weeks". The Project Manager decides the first assessment date to suit the parties and following assessments are carried out within the assessment intervals.
- The interest rate on late payments is stated.
- The period in which payments are made (if not three weeks).
- The share percentages (if Option C or D is used).
- The exchange rates (if Option C, D, E or F is used).

SECTION 6 – COMPENSATION EVENTS

- There are entries within this section in respect of the weather, the place where weather is to be recorded (weather station, airport, etc.), the weather measurements to be recorded, the default measurements being cumulative monthly rainfall, the number of days with rainfall more than 5mm, minimum air temperature less than 0 degrees Celsius, and snow lying at the designated time of day. The supplier of the weather measurements, the place where weather measurements are recorded, and where they are available from are also stated.

For some isolated site locations where no recorded data may be available, assumed values may be inserted.

- The value engineering percentage is 50% by default unless another percentage is stated.
- The method of measurement (if Option B or D is used).
- Additional compensation events.

SECTION 7 – TITLE

This section is not included within the Contract Data as no entries are required.

SECTION 8 – RISKS AND INSURANCE

- Any additional Client's liabilities are stated.
- The minimum limit of indemnity for third party public liability and the Contractor's liability for its own employees is stated within this section.

Contract Data Part 1 then contains a series of optional statements which are completed where appropriate:

- If the Client is to provide Plant and Materials, there is provision for insurance of the works to include any loss or damage to such Plant and Materials.
- The Contractor provides the insurances stated within the Insurance Table, except any insurances which the Client is to provide which are stated in Contract Data Part 1, which lists what the insurance is to cover, the amount of cover and the deductibles (otherwise known as "excesses"), the amounts to be paid by the insuring party, often before the insurance company pays.

Contract Data Part 1 then contains a series of entries related to resolving and avoiding disputes:

- The tribunal is stated.
- If the tribunal has been identified as arbitration, the Client must identify the arbitration procedure, the place where any arbitration is to be held, who will choose an arbitrator if either the parties cannot agree a choice, or if the arbitration procedure does not state who selects an arbitrator.
- The Senior Representatives of the Client are stated.
- The name and contact details of the Adjudicator are stated.
- With regard to dispute resolution, the Adjudicator nominating body is named within this section, normally a professional institution, and then the tribunal is named as either arbitration or legal proceedings.

If Option W3 is used:

- The number of members of the Dispute Avoidance Board is stated (either one or three members).
- The Client's nomination for the Dispute Avoidance Board is stated, also the frequency of visits to Site by the Dispute Avoidance Board, and the Dispute Avoidance Board nominating body.
- Finally, dependent on which secondary options are selected, the Client enters details for the relevant entry in the Contract Data.

Part 2

Part 2 is completed by the Contractor and includes:

- the name and contact details of the Contractor,
- the percentage for Fee,
- names of key people and documents submitted with the tender, and also
- any Scope provided by the contractor,
- details of any programme submission,
- details of the pricing document, i.e. activity schedule or bill of quantities,
- names of Senior Representatives,
- any entries in respect of Secondary Options.

11.6 EU Procurement Directives

Whilst not wishing this book to be geographically specific, it is worth some consideration of the European Union (EU) procurement directives that apply to all EU members' procurement within the public sector, above minimum monetary thresholds which are reviewed on a regular basis, and subject to EU-wide principles of non-discrimination, equal treatment and transparency. The regulations affect government departments, local authorities and health authorities, and also utilities companies operating in the energy, water and transport sectors.

The purpose of the procurement rules is to open up the public procurement market and to ensure the free movement of supplies, services and works within the EU.

The rules are enforced by the Members States' Courts and the European Court of Justice and require that all public procurement must be based on value for money, defined as *"the optimum combination of whole-life cost and quality to meet the user's requirement"*, which should be achieved through competition, unless there are compelling reasons to the contrary.

In addition to the EU Member States, the benefits of the EU public procurement rules also apply to a number of other countries outside Europe because of an international agreement negotiated by the World Trade Organisation (WTO), entitled the "Government Procurement Agreement" (GPA).

The EU Procurement Directives set out the legal framework for public procurement when public authorities and utilities seek to acquire supplies, services, or works (e.g. civil engineering or building), procedures which must be followed before awarding a contract when its value exceeds set thresholds which are reviewed on a regular basis, unless the contract qualifies for a specific exclusion, for example on grounds of national security.

The Directives require contracting authorities to provide details of their proposed procurement in a prescribed format, which are then published in the Official Journal of the European Union (OJEU).

All companies replying to an OJEU advertisement have an equal opportunity to express interest in being considered for tendering. As in all tendering exercises, clients must ensure that those companies selected to tender receive exactly the same information on which to make their bid.

The notice may be issued electronically and this service may be provided through an intermediary organisation.

Generally contracts covered by the Regulations must be the subject of a call for competition by publishing a Contract Notice in the OJEU.

The OJEU Notice process has associated minimum timeframes for receipt of tenders or requests to participate, dependent on procurement procedure, from the date the contract notice is sent.

11.7 Receipt of tenders to contract award

It is critical that tenderers are given sufficient time in which to prepare and submit their tenders.

The time and place for the delivery of tenders needs to be clearly stated. It should also be made clear to all tenderers that late tender returns are invalid and should not be considered, unless all are given a further opportunity to re-tender. Preferably, late tenders should be returned to the tenderer unopened.

The tender documentation should also make clear the period over which a tender may remain open for acceptance. This would normally be 3 months, giving the Client time to assess the tenders and make the appropriate award.

Opening tenders

Tenders should not be opened until after the date and time for submission to avoid any suspicion of collusion. No tenderer should be allowed to alter the terms of a tender after the closing time.

The opening of tenders in public or in front of tenderers is not recommended, as it can lead to misunderstandings. If it must be done, the presence (but not the content) of any qualification should also be indicated.

After opening the tenders any supporting documents should be evaluated. Each tender is confidential and no details should be released to any other tenderer or to anyone with a financial interest in the tendering.

Tender analysis

By carrying out detailed checks pre-tender, when tenders are received, the evaluation criteria should be very straightforward as all the tenders should be capable of being accepted, so it will normally just include any further information gained from the bid itself:

- price
- contractor's ability to complete on time
- contractor's ability to achieve quality standards
- contractor's proposed team and team structure

Note that with design and build tenders, the comparison of bids is far more complex, as the tenders are not being judged on the basis that each has priced the same job!

There has been a trend in the past 10 to 15 years for tenders to be accepted on the basis of quality elements of the bid, rather than on price alone.

A tender analysis should be prepared showing details of each of the tenders submitted including:

- Price submitted
- Any exclusions or qualifications to the price submitted
- Any arithmetical errors
- Any technical issues such as design requirements

There should be clear guidance on how to deal with late, incomplete or qualified bids.

The tender assessment report should present a clear and logically reasoned case for acceptance of the recommended tender. If the inclusion of technical detail is essential, sufficient explanation – in plain language – should also be included.

The following may be included in the report:

- a tabular statement of the salient features of all the tenders received, e.g. the name of the tenderer, the tender sum, whether valid (received by due date, complying with instructions for tendering, etc.), and qualifying conditions.
- reference to any discussions that may have taken place with any tenderers regarding the tender.
- a concise summary of the findings following the examination and analysis of each tender, reasons for considering any tender invalid, discussion of the programme and methods of execution proposed, and reasons for considering any of these unsatisfactory.
- comments on any rates, which appear exceptionally high or low, and forecasts of the possible effects on the contract price in cases where there are large differences.

- a comparison of the recommended tender sum with the cost plan or budget.
- recommendation of the most acceptable tender.
- recommendations for dealing with any qualifications in the recommended tender prior to acceptance.

Only after a tender has been accepted, should unsuccessful tenderers be advised of the name and tender price of the successful tenderer.

REMEMBER – THE CHEAPEST PRICE DOES NOT ALWAYS MEAN THE CHEAPEST COST!

Notification of tender results

It is important that all competing contractors be informed of the results of tendering as soon as possible after the tender submission date. This particularly applies to those whose tenders are not successful or which are not being given further consideration, as it allows them to concentrate on opportunities with other clients.

Generally, a Client's representatives have no authority to accept tenders, as ultimately the contract is between the Client and the Contractor, so it is normal to recommend a choice to the Client with him making the final decision and informing the tenderers accordingly.

The Client's representative should submit a report, which lists the tenders received, comments on them and recommends a particular one for acceptance. The scope and detail of this report will vary according to the circumstances.

Awarding the contract

Once the tenders have been assessed and a decision made as to which tender is successful and the contract awarded, it is essential that the contract be awarded as soon as practicable, and also that unsuccessful tenderers be advised accordingly.

At this stage, it is useful to follow a checklist approach to ensure that all the pre-construction issues are covered.

These will include:

- Confirm order or issue letter of intent;
- Prepare contracts for signing;
- Agree key dates, sequence of works and programme, including integration with the project programme;
- Agree timing for issue of drawings or approval of design information;
- Confirm requirements for insurances;
- Agree or obtain client/client representative approval of subcontracts or packages where there is a contractual requirement to do so;

- Agree the provision of samples, sample panels and mock-ups to be submitted for client approval;
- Agree the facilities to be provided by the contractor, for example, cabins, stores, offloading, scaffolding, removing rubbish, etc;
- Agree dates for interim certificates and procedure for requesting payment;
- Agree notification requirements for instructions, variations, dayworks, etc;
- Sign contract.

There is no Standard Form of Agreement for completion and signature within the NEC4 contracts as the drafters of the contract state that parties enter into contracts in many different ways. Whilst this is probably true, it would have assisted the parties if there had been a standard Form of Agreement that they could use if they so wish.

The difficulty with not having a Form of Agreement within the Contract is that the parties have to write their own, and in that respect must be very cautious that all the documents that are intended to form the contract are included and stated on the Form of Agreement.

Following the report on tenders the next move is with the Client. Decisions taken on any matters which are not included in the contract documents, or which vary them, must be carefully recorded by means of an exchange of letters.

In order to satisfy the common law test of offer and acceptance, the acceptance of a tender must be unconditional. Thus a letter of acceptance is not the place to introduce additional conditions, like "Your tender is accepted subject to our obtaining the necessary finance within one month".

Such new conditions would be a counter-offer, cancelling the original offer and in turn requiring acceptance by the tenderer. While some negotiation between the parties is not unusual, eventually an unconditional acceptance of a current offer (or counter-offer) is needed.

Formalising the contract

The signing of a formal agreement only adds additional evidence to the contract, but is still desirable, particularly if there have been complex negotiations. It is important to ensure that the signed documents incorporate all the documentation on which tenders and the final offer and acceptance are based. At the time of signing of the agreement, the parties, together with any amendments, which may have been made to printed documents, should initial every page of all such other documents.

When arranging the signing of contract documents, a check should be made that the signatories have legal authority to make contracts on behalf of their organisation, and that the formalities required by the organisation's rules are followed. This arises particularly with companies, incorporated societies or statutory bodies; typically, they may require the placing of the organisation's seal and two authorised signatures.

Letter of acceptance of tender

This is a letter from the Client to the Contractor accepting the offer. If negotiations between Client and Contractor take place before a tender is finally accepted, copies of any other letters giving evidence of the conditions agreed to should also be incorporated in the documents. In order to form a contract, there must be a well-defined offer to which an unconditional acceptance has been given.

Index

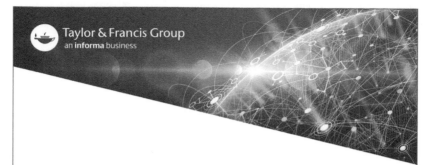